PATTERN RECOGNITION
Techniques and Applications

Related Oxford Titles in Electrical/Electronics and Communication Engineering

Allen and Holberg, *CMOS Analog Circuit Design*
Bobrow, *Fundamentals of Electrical Engineering, 2nd Edition*
Campbell, *The Science and Engineering of Microelectronic Fabrication, 2nd Edition*
Chen, *Digital Signal Processing*
Chen, *Linear System Theory and Design, 3rd Edition*
Chen, *System and Signal Analysis, 2nd Edition*
DeCarlo and Lin, *Linear Circuit Analysis, 2nd Edition*
Guru and Hiziroglu, *Electric Machinery & Transformers, 3rd Edition*
Islam, *Semiconductor Physics and Devices*
Krein, *Elements of Power Electronics*
Kuo, *Digital Control Systems, 3rd Edition*
Lathi, *Modern Digital and Analog Communications Systems, 3rd Edition*
Martin, *Digital Integrated Circuit Design*
Moorthi, *Power Electronics*
Nagsarkar and Sukhija, *Basic Electrical Engineering*
Ramakalyan, *Linear Circuits: Analysis and Synthesis*
Sadiku, *Elements of Electromagnetics, 3rd Edition*
Schaumann and Van Valkenburg, *Design of Analog Filters*
Sedra and Smith, *Microelectronic Circuits, 5th Edition*
Stefani, Savant, Shahian, and Hostetter, *Design of Feedback Control Systems, 3rd Edition*

PATTERN RECOGNITION

Techniques and Applications

RAJJAN SHINGHAL

Formerly, Professor, Concordia University,
Montreal, Canada

OXFORD
UNIVERSITY PRESS

OXFORD
UNIVERSITY PRESS

YMCA Library Building, Jai Singh Road, New Delhi 110001

Oxford University Press is a department of the University of Oxford.
It furthers the University's objective of excellence in research, scholarship,
and education by publishing worldwide in

Oxford   New York
Auckland   Cape Town   Dar es Salaam   Hong Kong   Karachi
Kuala Lumpur   Madrid   Melbourne   Mexico City   Nairobi
New Delhi   Shanghai   Taipei   Toronto

With offices in
Argentina   Austria   Brazil   Chile   Czech Republic   France   Greece
Guatemala   Hungary   Italy   Japan   Poland   Portugal   Singapore
South Korea   Switzerland   Thailand   Turkey   Ukraine   Vietnam

Oxford is a registered trade mark of Oxford University Press
in the UK and in certain other countries.

Published in India
by Oxford University Press

ISBN-13: 978-0-19-567685-3
ISBN-10: 0-19-567685-8

Typeset in Computer Modern
by Archetype, New Delhi 110002
Printed in India by Ram Book Binding House, New Delhi 110020
and published by Manzar Khan, Oxford University Press
YMCA Library Building, Jai Singh Road, New Delhi 110001

Preface

We often learn from examples: by looking at several roses, dahlias, pansies, orchids, daisies, petunias, we learn to recognize them on our own. We generalize on the flowers we saw, and then apply these generalizations to recognize flowers of these classes, which may include flowers we had not seen earlier. This illustrates pattern recognition in the domain of flowers, each flower being considered to be a pattern. Pattern recognition can be applied in other domains, such as optical character recognition, predicting the minerals in an area by studying the soil samples of that area, and assessing credit risk in a financial transaction.

About the book

This book focuses on procedures that can be implemented on a computer for pattern recognition in general. As a textbook, it is aimed at undergraduate and postgraduate students of computer science and electrical engineering.

Having myself once been an engineering student, and later having taught engineering students, I have found that, on seeing some theory, they also like to see how it is applied. The procedures presented show how the theory is applied. Moreover, it is shown how the procedures work on examples. In their applications, if students need to modify a procedure, they can do so, because they will know the underlying theory.

To understand the theory, the students should have the mathematical background required for a typical engineering or computer science degree: sets, probabilities, coordinate geometry, matrices, and calculus. To implement the procedures described, they should be fluent in at least one programming language. Moreover, they should know how to implement data structures such as lists and trees, and they should know about databases. Students should have studied these topics in other courses before they study this course. Programming languages have kept evolving over time, and I have no preference for any particular language. The procedures have accordingly been described in English pseudo-code so the students can implement them in a programming language of their choice. Since the procedures described can be applied to different domains, pattern-recognition practitioners working in industrial and research organizations, too, may find this book useful.

Content and Structure

Readers should certainly begin from the first chapter, which explains what pattern recognition is. The chapter then gives an overview of the material that follows in the rest of the book.

The second and third chapters explain how decision trees are used for classification. The fourth chapter discusses an evolutionary procedure that iteratively refines its learning, which can later be used for classification. The Bayes classifier is described in the fifth chapter. This classifier uses probabilities. The nearest neighbour classifier, based on the maxim 'Birds of a feather flock together', is in the sixth chapter. Then the seventh chapter discusses neural nets. Linear classifiers are presented in the eighth chapter. The ninth chapter describes a procedure to select the most useful attributes. Unsupervised learning, in the form of clustering, is described in the tenth chapter. The eleventh chapter presents a syntactic approach to pattern recognition. Closing remarks on the different classification procedures appear in the twelfth chapter.

The theory discussed in the chapters is applied to examples. Chapter summaries provide the highlights of the material covered. Programming and non-programming exercises are also given at the end of the chapters.

The book has three appendices. Appendix A reviews probabilities. The other two appendices describe projects that may be undertaken by students. The project in Appendix B is on breast-cancer prognostication, and that in Appendix C is on optical character recognition (OCR). An attached CD provides data files for the OCR project.

Acknowledgements

The Commissioning Editor of Oxford University Press (OUP) asked me to write this book. On my first draft, I received from OUP the comments of three reviewers, whose names I do not know. Incorporating their suggestions improved the manuscript. To get the manuscript ready to be published, and to rearrange some of the material, I was guided by OUP's editorial team.

Undoubtedly, there were others, too, on the OUP staff, whom I never got to know, but who worked on transforming the manuscript into a book. To all these people, I offer my sincere gratitude.

RAJJAN SHINGHAL

Contents

1

Learning to Recognize Patterns

1.1 Preview of Inductive Learning

Without doubt, we can instinctively distinguish a cat from a dog, regardless of whether we see the animal in a picture or in reality. We do not have to wait for the animal to bark or meow to recognize it. Think of your childhood when some grown-up pointed to an animal and told you that it was a cat and that some other animal was a dog. That grown-up need not have explained to you what made one animal a cat, and another a dog. Suppose each dog and cat is considered to be a pattern. Over time, you saw more patterns of cats and dogs and a stage came when no one had to recognize the animal for you. You were able to recognize the animal yourself. You had learnt from examples.

Recognizing an animal is the same as classifying the animal, that is, assigning the animal either to the class of cats or to the class of dogs. As is often done in pattern recognition, the phrases *recognize a pattern* and *classify a pattern* will be used synonymously in this book, depending on which phrase fits better in a given sentence.

The set of cats and dogs you saw initially was your *training set*— you trained yourself on the patterns of that set, each pattern being a *training pattern*. When you saw a cat or a dog that you had never seen before (it was not a training pattern) and tried to recognize it, then that pattern belonged to the *recall set*—you recalled your training, the pattern being known as a *recall pattern*. If we can recognize only the training patterns, but none of the recall patterns, then all we need to do is to memorize the training patterns. Such rote learning is not discussed in this book. Our aim is to learn how to recognize recall patterns.

From the training patterns of cats and dogs, you had intuitively developed some general rules to figure out what a cat is and what a dog is. Then, when you saw a recall pattern, you instinctively applied your rules to that pattern and decided whether it was a cat or a dog.

To develop rules that work well on different recall patterns, it is essential that the training set comprises a variety of cats and dogs. If the dogs in the training set are only beagles, dachshunds, greyhounds, poodles, and retrievers, then you might find it difficult to recognize the boxers and terriers of the recall set. The training and recall sets, though disjoint, should accordingly be taken from the same representative sampling.

As an exercise, list the things you observe before deciding whether a given animal is a cat or a dog. It would not be surprising if you find it difficult to draw up the list. So, although we can instinctively recognize cats and dogs, we might find it difficult to formalize the differences between them.

If we want to automate distinguishing a cat from a dog so that a computer is able to do so, we would first need to formalize the differences between the two classes of animals. To get an idea of how to automate this, let us, as an illustration, choose LENGTH as an *attribute* to measure the animal from its snout to the tip of its tail. We first set up a training set of cats and dogs. We then feed the value of LENGTH for each training pattern and the identity (cat or dog) of its corresponding class to a computer, in which a program—let us call it the *learning program*—analyses the values. Suppose the program observes that the LENGTH of every dog is more than 45 cm and the LENGTH of every cat is 45 cm or less (45 cm thus serves as a *threshold* value for LENGTH). Then the program formulates the following two rules:

> If the LENGTH of the animal is more than 45 cm,
> then the animal belongs to the class dog.

> If the LENGTH of the animal is 45 cm or less,
> then the animal belongs to the class cat.

Rules written in the If . . . then form, such as above, are known as *production rules* which we will shorten to *prules* (pronouncing the 'pru' as we do in 'prunes'). The expression between the 'If' and the 'then' of a prule is the *antecedent* of the prule, and the expression after the 'then' is the *consequent* of the prule. Prules represent a generalization, commonly known as *induction* (or *inductive learning*), done on the training patterns.

The prules generated by the learning program are stored in the computer memory. They constitute the *knowledge base* for recognizing patterns (cats and dogs). Once the knowledge base has been built, the *training phase* is over.

We now begin the *testing phase* to find out how good the knowledge base is in recognizing other cats and dogs. For this, we need another set of cats and dogs—the set ideally containing only recall patterns, although we may have a set containing a mixture of training and recall patterns. To test the knowledge base, the value of LENGTH for an animal is fed to the computer, in which a program called the *inference engine* checks for a prule whose antecedent is made true by the value of LENGTH. The engine then infers the consequent of that prule. The prule is said to have been *fired*. The consequent inferred is the output, which indicates the class of the animal. The procedure is repeated for the remaining patterns in the set.

From the computer's output, we may notice that some animals have been recognized correctly, and some incorrectly—that is, *misrecognized*—when a cat is identified as a dog, or vice versa. This happens because the set used in the testing phase could have cats whose length is more than 45 cm and dogs whose length is 45 cm or less. Such animals are exceptions to the prules in the knowledge base.

We may carry out the testing phase again after changing the threshold value (in this case 45 cm) for LENGTH in the prules, and yet misrecognize some of the animals. In fact, an ideal threshold value for LENGTH, which helps recognize all animals correctly, may not exist. To consider only one attribute (of LENGTH) is perhaps not enough to always correctly distinguish a dog from a cat. We may choose to have other attributes as well, such as COLOUR-OF-EYES, SHAPE-OF-HEAD, and FORM-OF-PAWS. For each attribute, we need to specify its possible values, which may be numeric as for LENGTH or non-numeric as for COLOUR-OF-EYES (black, brown, grey, etc.).

Since LENGTH has continuous values, we can say that LENGTH has infinite possible real values. If, however, we threshold LENGTH as we have done above, then we have quantized it into two possible values—either greater than the threshold value or less than or equal to the threshold value. It is in fact common practice to quantize numeric values into disjoint ranges. For instance, some numeric attributes may be quantized into these ranges: $[0, 10)$, $[10, 20)$, $[20, 30)$, ... , where, as is usual in mathematical notation, parentheses denote open intervals and square brackets denote closed intervals.

If an attribute has only one possible value, then every pattern will have the same value for that attribute regardless of the class of the pattern. Such an attribute will be useless for classifying a given pattern. This does not mean that every attribute with two or more possible values is useful for classifying patterns. For instance, the attribute AGE (in years) may have many possible values, but we can intuitively see that it will not be of much use in distinguishing a cat from a dog. For an attribute to be useful for classifying, it is necessary, but not sufficient, for it to have two or more possible values. There should, however, be at least one attribute; on which we can base our classification of patterns. So for the rest of this book, it will be assumed that there is at least one attribute for each pattern, and each attribute has at least two possible values.

It is certainly possible to take all training patterns from one class, but that is not challenging, because all recall patterns will then be assigned to that single class. For the training to be challenging, the training patterns should be taken from at least two classes. Although, in general, a pattern may belong to more than one class (a dachshund belongs to the class of dogs and also to the class of mammals), it will be assumed in this book that no pattern belongs to more than one class. In other words, the classes are disjoint, with each pattern belonging to exactly one class.

Let us select multivalued attributes A_1 to A_M, where $M \geq 1$, to recognize cats and dogs. The learning program can then be expected to formulate prules like these:

If A_1 has value$_1$,
and A_2 has value$_2$,
$\vdots$

and A_M has value$_M$,
then the animal belongs to the class dog.

If A_1 has value$'_1$,
and A_2 has value$'_2$
$\vdots$

and A_M has value$'_M$,
then the animal belongs to the class cat.

If TRUE,
then 'I am unable to recognize the animal'.

As we will see later, it is not necessary for each of the M attributes to be examined in the antecedent of each prule. The values of some of the attributes may not be needed to assign a pattern to a particular class. Each prule has up to M *conditions* in its antecedent, each condition examining the value of one attribute.

The last prule shown above is known as the *default prule*. Its antecedent, being always true, is a *tautology*. For some animal, if neither of the first two prules fires, then the third prule fires automatically and the computer responds by saying it is unable to recognize the animal. The computer is said to have *rejected* the animal. Either a human has to recognize the animal or the prules in the knowledge base have to be modified. Having a default prule ensures that no matter what attribute values are fed to the computer, it returns with a response, as is expected of a user-friendly computer.

The consequent of a default prule need not always indicate the rejection of a pattern. It is up to the designer of the knowledge base to decide what the consequent of a default prule should be. Suppose we notice that, in the training set, 60 per cent of the animals are dogs and 40 per cent cats. We may then assume that there are more dogs in the world. So if we come across a pattern that fires none of the first two prules, then the default prule consequent could say that the animal is a dog. A default prule with such a consequent can, however, be inappropriate. For instance, if a computer is fed a patient's symptoms with attribute values such as the rate of heart beat, blood pressure, and white-blood cell count, and the computer has to respond by diagnosing the patient's disease, then it is better for a default prule to reject the symptoms (say it does not know what disease the patient has) rather than to say that the patient has a disease which happens to be most prevalent. The latter could be a misdiagnosis, which in turn could lead to a wrong, perhaps harmful, therapy for the patient. Unless otherwise specified, we will assume that the consequent of a default prule indicates the rejection of a given pattern.

The above prules have been written in informal English and will continue to be written so in this book. That will make them easy to read and understand. For storing them in computer memory and making inferences, one should use a precise language, such as that used in predicate logic.

1.2   Bigotry and Inductive Learning

Suppose you have three young families from Tambolia (an imaginary country) living in your neighbourhood, and it is observed that they keep their houses dirty. By induction on the training set of three Tambolian families, one native resident of the neighbourhood may develop the rule that all families from Tambolia keep their houses dirty. Again by induction, another native resident may develop the rule that all *young* families from Tambolia keep their houses dirty. The two rules can be formulated as the following prules:

> If the family is from Tambolia,
> then it keeps its house dirty.

> If the family is from Tambolia,
> and the family is young,
> then it keeps its house dirty.

The first prule will fire for all Tambolian families, whereas the second prule will fire only for young Tambolian families, which are a proper subset of all Tambolian families. So an elderly Tambolian family will keep its house dirty according to the first prule, but not according to the second.

The two native residents have learnt differently from the same training set of the three young Tambolian families. We may say that both these prules typify bigotry. Thus, one can learn bigotry by using induction, but then unlearn it, again by induction. Suppose another young Tambolian family moves into the neighbourhood and keeps its house clean. One would then realize that both the earlier prules need to be modified because it evident that not all young Tambolian families keep their houses dirty. Learning something by induction often consists of iteratively refining what we had learnt earlier, which could sometimes also mean unlearning what has been learnt earlier.

Bigotry can be learned by induction, but that does not mean induction is to be discouraged. On the contrary, it can be encouraged: doctors continuously improve their inductive learning over the years, as they see more and more patients (similar to adding more patterns to their training set), and that leads to a continuous improvement in their skill of treating patients. Just as a surgeon is expected to use his knife to remove a tumour, not slit a throat, induction is expected to be put to good use.

1.3   A Professor–Student Example

In this example, the objective is to distinguish professors from students by observing three of their attributes HABIT, EATS, and FOOTWEAR.

NAME of training pattern	Attributes			Class
	HABIT	**EATS**	**FOOTWEAR**	
T1	Gabby	Baked	Clogs	Student
T2	Gabby	Roasted	Sandals	Professor
T3	Gabby	Baked	Sandals	Student
T4	Quiet	Fried	Sandals	Professor
T5	Gabby	Fried	Clogs	Student
T6	Quiet	Baked	Sandals	Student
T7	Gabby	Fried	Sandals	Professor
T8	Quiet	Fried	Clogs	Student

FIGURE 1.1 Training set for recognizing professors and students. The possible attribute values are 'gabby' and 'quiet' for HABIT; 'baked', 'fried', and 'roasted' for EATS; and 'clogs' and 'sandals' for FOOTWEAR.

Figure 1.1 provides a training set[1] of eight patterns of professors and students. The caption of the figure lists the possible values for each of these three attributes. The figure is, in fact, a table, but all tables, diagrams, and any textual aides will be referred to as figures in this book, for that puts all of them in a common sequence, thus making it easier to search for them.

Study Fig. 1.1, and by intuition, come up with a set of prules for each of the two classes. This is similar to what a learning program does except that it does not use intuition. Now jot down justifications for the prules developed. Apply your prules to infer the classes of the four recall patterns given in Fig. 1.2. Figures 1.1 and 1.2 will be referred to at many places in the book, hence it is advised that one becomes well familiar with them.

Now you may ask seven friends to infer the classes of the recall patterns based on their intuitive study of the training set. Do not be surprised if their classifications of the recall patterns are not the same as yours. If a

---

[1] I got the idea of this training set while reading an example on lions in P. Clark 1990, 'Machine Learning: Techniques and Recent Developments', in A.R. Mirzai, (ed.), *Artificial Intelligence: Concepts and Application in Engineering,* MIT Press, Cambridge, Massachusetts, pp. 65–93.

NAME of recall pattern	Attributes			Class
	HABIT	EATS	FOOTWEAR	
R1	Quiet	Baked	Clogs	
R2	Quiet	Roasted	Sandals	
R3	Gabby	Roasted	Clogs	
R4	Quiet	Roasted	Clogs	

FIGURE 1.2   Recall set for recognizing professors and students. The training set is given in Fig. 1.1.

person's classifications are different from yours, then what that person learnt by induction from the training set was different from what you learnt by induction from the same training set (remember how the two native residents learnt differently about the Tambolian families in Section 1.2).

1.4 A Specific-to-General Procedure

The antecedent of a prule is said to *cover* a pattern if, and only if, the attribute values of the pattern make the antecedent true. Antecedent α_1 of a prule is defined to be *more specific* than antecedent α_2 of another prule if, and only if, the patterns covered by α_1 are a proper subset of the patterns covered by α_2. In other words, every pattern covered by α_1 is covered by α_2, but not conversely. In the two prules given about Tambolian families in Section 1.2, you will notice that the antecedent of the second prule is more specific than the antecedent of the first prule. Antecedent α_2 is said to be *more general* than antecedent α_1 if, and only if, α_1 is more specific than α_2. Thus, the first prule about Tambolian families is more general than the second prule.

The Specific-to-General (abbreviated as SpecToGen) procedure to generate prules for recognizing patterns of a particular class from a given training set starts with a prule whose antecedent is so specific that it covers no training patterns. The antecedent is then gradually generalized so that it covers the training patterns of the class. This procedure can then be repeated for each of the other classes of the training set. This procedure is as follows.

1. Let $M \geq 1$ be the number of attributes A_1 to A_M.

2. Let $m \geq 1$ be the number of classes C_1 to C_m.

3. Initialize a set ϕ to empty. This set will eventually contain the prules for the m classes.

4. For $k = 1, 2, \ldots, m$ do steps 4.1 to 4.3.

   4.1. Initialize the antecedent of a prule to its most specific instance of a conjunction of M conditions, each condition being FALSE. It can be expressed as $\langle \mathrm{FALSE}_1, \mathrm{FALSE}_2, \ldots, \mathrm{FALSE}_M \rangle$.

   4.2. Process each training pattern in class C_k according to step 4.2.1.

      4.2.1. For $i = 1, 2, \ldots, M$, do step 4.2.1.1 (after this step is over, the antecedent will cover the training pattern being processed).

         4.2.1.1. If the value of attribute A_i of the pattern does not make the ith condition of the antecedent true, then replace the ith condition by its more general form so that the value of A_i makes it true.

   4.3. If a condition in the antecedent contains a disjunction of all possible values of an attribute, then delete that condition from the antecedent, since the condition will always be true. With the antecedent, create a prule whose consequent says that the pattern is in class C_k. Put the prule in set ϕ.

5. Return with the prules in set ϕ.

To illustrate this procedure, let us apply the SpecToGen procedure to obtain prules for the professor–student training set shown in Fig. 1.1, for which $M = 3$ because of the three attributes (HABIT, EATS, and FOOTWEAR) and $m = 2$ because of the two classes (professor and student). Let us first obtain prules for the class of professors. We begin with the antecedent

$$\langle \mathrm{FALSE}_1, \mathrm{FALSE}_2, \mathrm{FALSE}_3 \rangle$$

One by one, we will process the training patterns that belong to the class of professors. Processing a pattern consists of generalizing the antecedent so that it covers the pattern. The first pattern we process is T2; the antecedent becomes

$$\mathrm{T2:} \langle \mathrm{HABIT} = \mathrm{gabby}, \mathrm{EATS} = \mathrm{roasted}, \mathrm{FOOTWEAR} = \mathrm{sandals} \rangle$$

For ease of understanding, the name of the training pattern (for example, T_2) will be written before the antecedent to indicate what the antecedent contains after that pattern has been processed. The antecedent covers pattern T2. We next process pattern T4. To ensure that the antecedent covers T4, we generalize the antecedent to

$$T4:\langle HABIT = gabby\ or\ quiet, EATS = roasted\ or\ fried,$$
$$FOOTWEAR = sandals\rangle$$

The antecedent already covers pattern T7, and hence it need not be generalized any further. So the antecedent remains as it was:

$$T7:\langle HABIT = gabby\ or\ quiet, EATS = roasted\ or\ fried,$$
$$FOOTWEAR = sandals\rangle$$

There are no more training patterns remaining from the class of professors. In the first condition of the above antecedent, we notice that HABIT can take either of its two possible values. The condition containing HABIT will always be true. Accordingly, HABIT need not be examined, and we can delete the first condition from the antecedent. The antecedent thus reduces to

$$\langle EATS = roasted\ or\ fried, FOOTWEAR = sandals\rangle$$

The prule developed from this antecedent can be written as

If EATS = roasted or fried,
and FOOTWEAR = sandals,
then the pattern belongs to the class professor.

It is customary to split a prule containing an 'or' in the condition of an antecedent into two or more prules. From the distributivity law of mathematical logic, the above prule can be rewritten as the two prules SG1 and SG2 shown in Fig. 1.3.

Let us now obtain the prules for the class of students. We again begin with the antecedent

$$\langle FALSE_1, FALSE_2, FALSE_3\rangle$$

SG1: If EATS = roasted,

and FOOTWEAR = sandals,

then the pattern belongs to the class professor.

SG2: If EATS = fried,
and FOOTWEAR = sandals,

then the pattern belongs to the class professor.

SG3: If EATS = baked,

then the pattern belongs to the class student.

SG4: If EATS = fried,

then the pattern belongs to the class student.

SG5: If TRUE,

then 'I am unable to classify the pattern'.

FIGURE 1.3 Knowledge base comprising prules SG1 to SG5, obtained by the application of the SpecToGen procedure to the professor–student training set shown in Fig. 1.1.

One by one, let us process the training patterns belonging to the class of students. The antecedents obtained after processing these patterns are

T1:$\langle$HABIT = gabby, EATS = baked, FOOTWEAR = clogs$\rangle$

T3:$\langle$HABIT = gabby, EATS = baked, FOOTWEAR = clogs or sandals$\rangle$

T5:$\langle$HABIT = gabby, EATS = baked or fried,

  FOOTWEAR = clogs or sandals$\rangle$

T6:$\langle$HABIT = gabby or quiet, EATS = baked or fried,

  FOOTWEAR = clogs or sandals$\rangle$

T8:$\langle$HABIT = gabby or quiet, EATS = baked or fried,

  FOOTWEAR = clogs or sandals$\rangle$

In the first condition of last antecedent T8, we notice that HABIT can take either of its two possible values; that is, this condition will always be true. Accordingly, HABIT need not be examined, and we can delete the first condition from the antecedent. Similarly, in the third condition,

FOOTWEAR can take either of its two possible values, and we can accordingly delete this condition. The antecedent thus reduces to

$$\langle \text{EATS} = \text{baked or fried} \rangle$$

The prule developed from the above antecedent is then

If EATS = baked or fried,
then the pattern belongs to the class student.

Because of the 'or' in the condition of the antecedent of the above prule, it can be rewritten as the two prules SG3 and SG4 shown in Fig. 1.3. None of the prules SG1 to SG4 have HABIT in their antecedents. To prules SG1 to SG4, the default prule SG5 has been added, as mentioned in Section 1.1.

Using prules SG1 to SG5, the classifications of the eight training patterns given in Fig. 1.1 and the four recall patterns given in Fig. 1.2 are shown in Fig. 1.4. Two of the four recall patterns have been rejected. In addition, the training patterns T4 and T7 have been classified under the class professor by firing prule SG2 and under the class student by firing prule SG4. These two training patterns can also be considered as rejected. The inference engine, however, is often so designed that if a general and a specific prule can fire, then only the specific prule is fired. With this design of the inference engine, SG2 fires, not SG4, since SG2 is more specific than SG4. Patterns T4 and T7 will then be classified under the class professor.

In general, the number of prules in a knowledge base may be large. To reduce this number, we should replace multiple occurrences of a prule by a single occurrence. Moreover, we can temporarily remove a prule from the knowledge base and then test the knowledge base on the training set. If the number of training patterns recognized correctly after the prule is removed is greater than or equal to the training patterns recognized correctly before it was removed, then we can delete the prule permanently from the knowledge base. We can repeat this test for each prule. Deletion of redundant prules will result in a smaller knowledge base, which should increase the efficiency whenever the knowledge base is used. This method of reducing the size of the knowledge base can always be applied no matter which procedure we use to develop the knowledge base.

NAME of pattern	Attributes			Prule fired	Classi-fication
	HABIT	EATS	FOOTWEAR		
T1	Gabby	Baked	Clogs	SG3	Student
T2	Gabby	Roasted	Sandals	SG1	Professor
T3	Gabby	Baked	Sandals	SG3	Student
T4	Quiet	Fried	Sandals	SG2	Professor
				SG4	Student
T5	Gabby	Fried	Clogs	SG4	Student
T6	Quiet	Baked	Sandals	SG3	Student
T7	Gabby	Fried	Sandals	SG2	Professor
				SG4	Student
T8	Quiet	Fried	Clogs	SG4	Student
R1	Quiet	Baked	Clogs	SG3	Student
R2	Quiet	Roasted	Sandals	SG1	Professor
R3	Gabby	Roasted	Clogs	SG5	Rejection
R4	Quiet	Roasted	Clogs	SG5	Rejection

FIGURE 1.4 Classifications of the professor–student training patterns (T1 to T8 given in Fig. 1.1) and the recall patterns (R1 to R4 given in Fig. 1.2), using prules SG1 to SG5 given in Fig. 1.3, obtained by the SpecToGen procedure. The recall patterns R3 and R4 were rejected because none of the prules SG1 to SG4 fired for these two patterns, and hence the default prule SG5 fired.

1.5   A General-to-Specific Procedure

The General-to-Specific (abbreviated as GenToSpec) procedure is a converse to the SpecToGen procedure of Section 1.4. To generate prules for recognizing a particular class from a given training set, the GenToSpec procedure starts with a prule whose antecedent is so general that it covers all training patterns. The antecedent is gradually made more specific so that it covers only the training patterns of that class. The procedure is then repeated for each class of the training set. The procedure is as follows.

1. Let $m \geq 1$ be the number of classes C_1 to C_m.
2. For $k = 1, 2, \ldots, m$, do steps 2.1 to 2.3.

   2.1. Let the training patterns in class C_k be known as *positive patterns* and the training patterns not in C_k be known as *negative patterns*.

   2.2. Initialize a set ϕ_k to empty. This set will eventually contain the prules that have class C_k in the consequent, that is, the consequent of each prule will classify the pattern in class C_k.

   2.3. For each positive pattern $T^\oplus$ that is not covered by the antecedent of any of the prules in set ϕ_k (pattern $T^\oplus$ is called the *seed* for this iteration), do steps 2.3.1 to 2.3.5.

      2.3.1. Initialize a set η to empty. This set will eventually contain antecedents that cover seed $T^\oplus$, but do not cover any of the negative patterns.

      2.3.2. Initialize the antecedent of a prule to its most general instance such that it covers all training patterns. It is expressed as $\langle \text{TRUE} \rangle$. Put the antecedent in set η.

      2.3.3. Process each negative pattern $T^\ominus$ according to step 2.3.3.1 (after this step, none of the antecedents in set η will cover the processed pattern $T^\ominus$).

         2.3.3.1. For each antecedent α in η, do step 2.3.3.1.1.

            2.3.3.1.1. If the antecedent α covers $T^\ominus$, then replace α in set η by all possible, more specific antecedents that cover $T^\oplus$, but do not cover $T^\ominus$.

      2.3.4. Initialize a set ρ to empty. Heuristically, select antecedents from η and put them in ρ. A commonly used heuristic for selection considers two possible situations. In the first situation, there exist subsets of antecedents, such that the antecedents in each subset cover a future seed not already covered by any of the antecedents of the prules in ϕ_k. From each such subset, select the antecedent with the fewest conditions—select arbitrarily from two or more antecedents that have the fewest conditions. If the first situation does not exist, then the second situation exists, in which case, select the antecedent with the fewest conditions—again, select arbitrarily from two or more antecedents that have the fewest conditions. The fewer the conditions in an antecedent, the quicker it will be to check whether the antecedent covers a given pattern. This is the *minimum description length* principle.

2.3.5. With each antecedent in set ρ, create a prule with a consequent that classifies the pattern in class C_k. Put the prule in set ϕ_k.

3. Return with the prules in sets ϕ_1 to ϕ_m.

Let us apply the GenToSpec procedure to obtain prules for the professor–student training set shown in Fig. 1.1, for which $m = 2$ because of the two classes (professors and students). Let us first obtain prules for the class of professors. For this class, the positive training patterns are T2, T4, and T7 and the negative training patterns are T1, T3, T5, T6, and T8.

We begin by making pattern T2 the seed and initializing set η to empty. Next, we put the antecedent $\langle \text{TRUE} \rangle$, which covers all the training patterns, into the set η. Now, one by one, process the negative patterns. Processing a negative pattern consists of making the antecedents in η more specific, so that they no longer cover the negative pattern, but continue to cover the seed. The first negative pattern we process is T1. To prevent the antecedent from covering T1, and yet cover seed T2, we replace the antecedent in η by the following two antecedents:

$$\text{T2}_1 : \langle \text{EATS} = \text{roasted} \rangle$$

$$\text{T2}_2 : \langle \text{FOOTWEAR} = \text{sandals} \rangle$$

For ease of understanding, each antecedent is given a name consisting of the name of the seed and a subscript to help identify the antecedent individually.

We next process the negative pattern T3. Antecedent T2_1 does not cover T3 and hence we do not replace it. Antecedent T2_2, however, covers pattern T3. To prevent it from covering T3, and yet cover the seed T2, it is replaced by antecedent T2_{21} (a subscript ij in the name of an antecedent indicates that it is the jth antecedent obtained by replacing an antecedent whose name had the subscript i). The set η now contains the following antecedents:

$$T2_1: \langle \text{EATS} = \text{roasted} \rangle$$

$$T2_{21}: \langle \text{EATS} = \text{roasted}, \text{FOOTWEAR} = \text{sandals} \rangle$$

The above antecedents do not cover any of the remaining negative patterns T5, T6, and T8. So, when we process these patterns, the antecedents are not replaced. We are now in step 2.3.4 of the GenToSpec procedure, and we have to select one or more antecedents from those given above. Neither of the antecedents cover the future seeds T4 and T7. So we are in the second situation described in the step. We select antecedent $T2_1$ because it has only one condition, whereas $T2_{21}$ has two conditions. The prule developed from antecedent $T2_1$ is GS1 shown in Fig. 1.5.

GS1: If EATS = roasted,
then the pattern belongs to the class professor.

GS2: If EATS = fried,
and FOOTWEAR = sandals,
then the pattern belongs to the class professor.

GS3: If EATS = baked,
then the pattern belongs to the class student.

GS4: If FOOTWEAR = clogs,
then the pattern belongs to the class student.

GS5: If TRUE,
then 'I am unable to classify the pattern'.

FIGURE 1.5 Knowledge base comprising prules GS1 to GS5, obtained by the application of the GenToSpec procedure from the professor–student training set given in Fig. 1.1. Comparing these with the prules given in Fig. 1.3, we see that the antecedent of GS1 is more general than that of SG1. Moreover, GS2 is identical to SG2, and GS3 is identical to SG3. The antecedent of GS4 has no relationship with that of SG4.

Next, we make the positive pattern T4 the seed and re-initialize set η to empty. We then put the antecedent $\langle \text{TRUE} \rangle$, which covers all the training patterns, into set η. After processing negative pattern T1, set η contains the following antecedents:

$\quad$ $T4_1 : \langle \text{HABIT} = \text{quiet} \rangle$

$\quad$ $T4_2 : \langle \text{EATS} = \text{fried} \rangle$

$\quad$ $T4_3 : \langle \text{FOOTWEAR} = \text{sandals} \rangle$

We now process the negative pattern T3. Antecedents $T4_1$ and $T4_2$ do not cover T3 and hence they are not replaced. Antecedent $T4_3$, however, covers T3. It can be prevented from covering T3 in two ways, so it is replaced by antecedent $T4_{31}$ and $T4_{32}$. Then, set η contains the antecedents

$\quad$ $T4_1 : \langle \text{HABIT} = \text{quiet} \rangle$

$\quad$ $T4_2 : \langle \text{EATS} = \text{fried} \rangle$

$\quad$ $T4_{31} : \langle \text{HABIT} = \text{quiet}, \text{FOOTWEAR} = \text{sandals} \rangle$

$\quad$ $T4_{32} : \langle \text{EATS} = \text{fried}, \text{FOOTWEAR} = \text{sandals} \rangle$

We have now understood how processing each negative pattern affects the antecedents in set η. So, shortening the explanation, given below are the contents of η after processing the remaining negative patterns.

After processing the negative pattern T5, the antecedents in η are

$\quad$ $T4_1 : \langle \text{HABIT} = \text{quiet} \rangle$

$\quad$ $T4_{21} : \langle \text{HABIT} = \text{quiet}, \text{EATS} = \text{fried} \rangle$

$\quad$ $T4_{22} : \langle \text{EATS} = \text{fried}, \text{FOOTWEAR} = \text{sandals} \rangle$

$\quad$ $T4_{31} : \langle \text{HABIT} = \text{quiet}, \text{FOOTWEAR} = \text{sandals} \rangle$

$\quad$ $T4_{32} : \langle \text{EATS} = \text{fried}, \text{FOOTWEAR} = \text{sandals} \rangle$

The antecedents in η after processing the negative pattern T6 are

$\quad$ $T4_{11} : \langle \text{HABIT} = \text{quiet}, \text{EATS} = \text{fried} \rangle$

$\quad$ $T4_{21} : \langle \text{HABIT} = \text{quiet}, \text{EATS} = \text{fried} \rangle$

$\quad$ $T4_{22} : \langle \text{EATS} = \text{fried}, \text{FOOTWEAR} = \text{sandals} \rangle$

$\quad$ $T4_{311} : \langle \text{HABIT} = \text{quiet}, \text{EATS} = \text{fried}, \text{FOOTWEAR} = \text{sandals} \rangle$

$\quad$ $T4_{32} : \langle \text{EATS} = \text{fried}, \text{FOOTWEAR} = \text{sandals} \rangle$

The last negative pattern to process is T8. After processing it, the antecedents in η are

$$T4_{111} : \langle HABIT = quiet, EATS = fried, FOOTWEAR = sandals \rangle$$

$$T4_{211} : \langle HABIT = quiet, EATS = fried, FOOTWEAR = sandals \rangle$$

$$T4_{221} : \langle HABIT = quiet, EATS = fried, FOOTWEAR = sandals \rangle$$

$$T4_{311} : \langle HABIT = quiet, EATS = fried, FOOTWEAR = sandals \rangle$$

$$T4_{32} : \langle EATS = fried, FOOTWEAR = sandals \rangle$$

We are now in step 2.3.4 of the GenToSpec procedure, and we have to select one or more antecedents from those given above. We are in the first situation described in the step, wherein there exists an antecedent $T4_{32}$, which covers the future seed T7 (the other antecedents do not cover T7). So we select $T4_{32}$, and the prule GS2 developed from this antecedent is shown in Fig. 1.5. We need not use pattern T7 as a seed because it is already covered by the antecedent of the prule GS2.

Having developed prules for the class of professors, let us now develop prules for the class of students. For this class, the positive patterns are T1, T3, T5, T6, and T8 and the negative patterns are T2, T4, and T7.

We now make the positive pattern T1 the seed and initialize set η to empty. We next put the antecedent $\langle TRUE \rangle$, which covers all the training patterns, into η. After processing the negative pattern T2, η contains the following antecedents:

$$T1_1 : \langle EATS = baked \rangle$$

$$T1_2 : \langle FOOTWEAR = clogs \rangle$$

These antecedents do not change when we process the negative patterns T4 and T7, since neither pattern is covered by either of the above antecedents. According to step 2.3.4 of the GenToSpec procedure, we have to select one or more antecedents from those given above. We are in the first situation described in the step: antecedent $T1_1$ covers the future seeds T3 and T6 and antecedent $T1_2$ covers the future seeds T5 and T8. Hence, we select both of these antecedents. Prules GS3 and GS4 developed from these antecedents are shown in Fig. 1.5. We need not use the patterns T3, T5, T6, and T8 as seeds because they are already covered by the antecedents of prules GS3 and GS4. To prules GS1 to GS4 given in Fig. 1.5, the default prule GS5 has been added, as mentioned in Section 1.1.

Using prules GS1 to GS5, the classifications of the eight training patterns given in Fig. 1.1 and four recall patterns given in Fig. 1.2 are shown in Fig. 1.6. All training patterns are classified correctly. Recall patterns R3 and R4 are classified under the class professor by firing prule GS1, and under the class student by firing prule GS4. These two patterns are thus rejected.

The recall patterns R3 and R4 have been rejected by the prules of both the SpecToGen and the GenToSpec procedures, but for different reasons (Figs 1.4 and 1.6). None of the SpecToGen prules SG1 to SG4 fired for the two patterns, and hence the default prule SG5 fired. For the GenToSpec

| NAME of pattern | Attributes | | | Prule fired | Classi- fication |
	HABIT	EATS	FOOTWEAR		
T1	Gabby	Baked	Clogs	GS3 or GS4	Student
T2	Gabby	Roasted	Sandals	GS1	Professor
T3	Gabby	Baked	Sandals	GS3	Student
T4	Quiet	Fried	Sandals	GS2	Professor
T5	Gabby	Fried	Clogs	GS4	Student
T6	Quiet	Baked	Sandals	GS3	Student
T7	Gabby	Fried	Sandals	GS2	Professor
T8	Quiet	Fried	Clogs	GS4	Student
R1	Quiet	Baked	Clogs	GS3 or GS4	Student
R2	Quiet	Roasted	Sandals	GS1	Professor
R3	Gabby	Roasted	Clogs	GS1	Professor
				GS4	Student
R4	Quiet	Roasted	Clogs	GS1	Professor
				GS4	Student

FIGURE 1.6   Classifications of the professor–student training patterns (T1 to T8 given in Fig. 1.1) and the recall patterns (R1 to R4 given in Fig. 1.2), using prules GS1 to GS5 shown in Fig. 1.5, obtained by the GenToSpec procedure.

prules, however, each of the two patterns caused more than one prule (GS1 and GS4) to fire, which gave different classifications.

The two knowledge bases (Figs 1.3 and 1.5) we have developed from the professor–student training set given in Fig. 1.1 are not the only possible ones, as will be evident in the later chapters.

1.6 Overview

This section is an overview of the procedures given in this book so that one can choose to read the procedures of interest.

Section 1.1 is an introduction to the concept of pattern recognition. There exist different procedures for classification, and not all of them provide the same results for the given training and recall sets, just as different people may give different classifications for the recall set given in Fig. 1.2.

In inductive learning, one learns by generalizing on a given set of training patterns. This learning can be applied to recall patterns, those on which one has not been trained. Inductive learning can sometimes lead to bigotry.

Section 1.2 differentiates inductive learning from bigotry. A professor–student example has been described in Section 1.3. Understanding this example needs no specialized domain knowledge. A reference to this example will appear in later chapters as other procedures are applied to it. The aim is to learn the underlying principles of the procedures from this example and then adapt these principles to the domain of your interest, be it medical diagnosis, assessing credit risk from an individual's background, recognizing military installations from aerial photographs, geological exploration from soil samples, fingerprint identification, weather prediction from satellite photographs, speech recognition, signature identification, optical character recognition, or even recognizing cats and dogs for that matter.

Two procedures—SpecToGen and GenToSpec—for inductive learning have been described in Sections 1.4 and 1.5 using the professor–student example. The procedures have been applied to some training and recall patterns, which distinguish professors from students. The two procedures give different results, illustrating that different procedures can infer differently from the same set of training patterns.

All procedures have been presented in pseudocode, a mixture of English and mathematics. The focus is more on the understandability of the pseudocode than on its optimality. Once the pseudocode has been understood, it can be coded optimally in a programming language of the reader's choice. What the British poet Alfred Lord Tennyson (1809–1892) said about brooks and men can be adapted to our procedures by saying that programming languages may come and go, but pseudocode will stay forever.

One commonly adopted procedure needs a data structure called a tree to decide which class a given pattern belongs to. The decision tree procedure is explained in Chapters 2 and 3. Reading Chapter 2 before reading Chapter 3 is a must.

If decision trees do not interest the readers, they can move on to Chapter 4: it discusses an evolutionary procedure in which we gradually refine our learning. This procedure falls in the category of genetic algorithms.

Chapter 5, which describes the Bayes classifier, bases learning on probability. Probabilities are estimated from a training set and then used to classify recall patterns.

Chapter 6 deals with patterns belonging to the same class as its nearest neighbours.

Chapter 7 discusses neural nets. If the other procedures can be said to simulate the mind, then neural nets simulate the brain.

Linear classifiers are presented in Chapter 8. These can be applied when patterns of different classes can be separated by a hyperplane in the multi-dimensional attribute space. Even when classes are not linearly separable, a classifier can be trained in a higher dimensional attribute space.

Classification is affected by the attributes we use. Chapter 9 describes how to select attributes that are most useful for classification.

The procedures described in Chapters 1 to 8 implement what is known as 'supervised learning'. At the time of learning from training patterns, the class of each pattern is known. This information about the classes of the patterns can be said to supervise our learning. In 'unsupervised learning', the classes of training patterns are not known. In this type of learning, patterns are organized into clusters, the patterns within one cluster being similar to one another, but dissimilar to the patterns of another cluster. Chapter 10 describes such clustering.

A syntactic approach to pattern recognition is presented in Chapter 11. It consists of determining whether a particular sequence of symbols has been produced by a given grammar.

No matter which chapters the readers choose to read after Chapter 1, they should read Chapter 12, which presents some closing remarks on the different procedures. After reading Chapter 12, the readers may want to go back and read chapters they had not read earlier.

Exercises

1. Design a database (relational, hierarchical, network, or some other) to store a set containing 10,000 patterns, each pattern having 20 attributes. For efficient training and testing on this set, you would need efficient access to the individual pattern. Design the access mechanism you will use.

2. By looking around me, suppose I have developed by induction the following rules.

    (a) Every wife complains about her husband.
    (b) Birds fly.
    (c) Children cry.
    (d) Women cry at weddings.
    (e) Men are rash drivers.
    (f) Left-handed people are handicapped.
    (g) People from Delhi are assertive.
    (h) People from Kolkata are polite.
    (i) People from Mumbai are formal.
    (j) People from Chennai talk fast.

    Although the word 'All' has not been written in the above rules, it is implied, as is usual in English. With reasoning, comment on the rules. Which do you think are true? Which false? Which bigoted? Which meaningless? Which humorous? (Charles II [1630–85], King of England [1660–85], is reported to have said that, in humour, we appear to hurt someone, but we do not hurt them.) Rewrite each rule (i) as a prule in semimathematical English, (ii) as an implication in predicate logic—omit this if you do not know predicate logic, and

(iii) as an if ... then statement in a programming language of your choice. To illustrate, the solutions to rule (a) are as follows.

  (i) If x is a wife, then x complains about her husband.

 (ii) $\forall x(\mathit{WIFE}(x)) \rightarrow \mathit{COMPLAINS\_ABOUT}(x, husband(x))$
     where the predicates WIFE and $\mathit{COMPLAINS\_ABOUT}$ and the function $husband$ are self-explanatory.

(iii) The solution will depend on the programming language you choose.

3. When is it acceptable to draw generalized conclusions by induction about other groups, races, religions, and nationalities? Would you find it acceptable if such conclusions were drawn about the group to which you belong? Discuss.

4. Do humans usually have considerable skills in learning by induction? Justify your answer.

5. Give an instance of something you learnt by induction and later found that you had learnt it wrong.

6. American scientist William A. Martin has said, 'You can never learn anything unless you almost know it already.' Do you agree with this in the context of learning by induction? Give reasons for your answer.

7. If a number of doctors are shown the same set of symptoms for a given patient, is it certain that all of them will prescribe the same therapy? Give reasons for your answer.

8. Do you know someone who, after seeing one doctor went to another doctor for a so-called second opinion? Why would someone want a second opinion? Justify.

9. In business schools, students are often taught from case studies. Through these case studies, they learn about specific companies, the actions they took to improve their business, and the results of their actions. Why are such case studies considered useful for student learning? Are these case studies training patterns? Discuss.

10. Have the cat–dog, Tambolian-family, and professor–student examples made the chapter easier to understand? Do you think such examples serve as training patterns? Discuss.

11.   In a programming language of your choice, implement the SpecToGen and GenToSpec procedures described in Sections 1.4 and 1.5. Use the program to develop knowledge bases for the professor–student training set shown in Fig. 1.1. Test the knowledge bases on the patterns given in Fig. 1.4 and record the classification results.

Learning Objectives

This chapter contains the following topics:

- review of the structure of a tree, its leaf and non-leaf nodes, and the arrangement of arcs between these nodes
- decision trees and their utility in deciding the class of a given pattern
- associating each non-leaf node with an attribute being examined and each leaf node with a class or with rejection
- a breadth-first procedure to build a decision tree from a given training set
- building different decision trees from the professor–student training set of the first chapter
- showing how different decision trees built from the same training set can classify a given recall pattern differently
- developing the mathematical model for the ratio of information gain an attribute provides when examined at a node of a decision tree
- building a decision tree in which the attribute selected to be examined at a node maximizes the ratio of information gain
- obtaining production rules from a given decision tree

2

Decision Trees: Basics

2.1   Trees

This section briefly discusses a data structure called a tree and the terminology used in trees, mainly to refresh the reader's memory and to remove any ambiguity about the terminology. It is advisable to periodically refer to Fig. 2.1, which illustrates a tree.

A *tree* is a set of one or more *nodes*. A node x_i, subscript i being either empty or a sequence of one or more non-negative integers, is joined to another node x_{ij} by an *arc* directed from x_i to x_{ij} (between two nodes there can be one arc at the most, and no arc is directed from a node to the node itself). Node x_{ij} is said to be a *child* of x_i, and x_i is the *parent* of x_{ij}. When subscript i is empty, an alternative way is not to write it; thus x_i may be written as x, and x_{ij} as x_j.

If a node x_i has $n \geq 2$ children $x_{i1}, x_{i2}, \ldots, x_{in}$, then the children are said to be one another's *siblings*. *Expanding* a node means generating its children. A node with no children is a *leaf* node; otherwise, it is a *non-leaf* node.

A node x_i is an *ancestor* of another node x_j if, and only if, either x_i is a parent of x_j, or x_i is a parent of an ancestor of x_j. No node is an ancestor

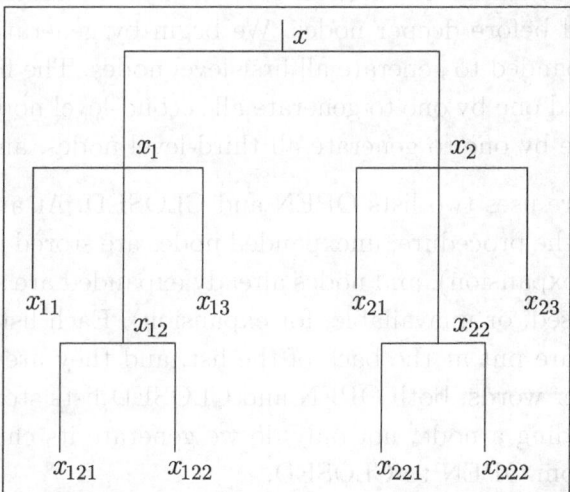

FIGURE 2.1   An illustration of a tree. The root node is x. Its two children are x_1 and x_2, node x_1 being the sibling of x_2 and vice versa. The root node is the ancestor of every other node in the tree. The leaf nodes are x_{11}, x_{13}, x_{21}, x_{23}, x_{121}, x_{122}, x_{221}, and x_{222}. The other nodes x, x_1, x_2, x_{12}, and x_{22} are non-leaf nodes. The level of root x is 0, and the level of nodes x_1 and x_2 is 1. The depth of the tree is 3.

of itself. A node x_j is a *descendant* of a node x_i if, and only if, x_i is an ancestor of x_j. The *root* node is the ancestor of all other nodes in the tree. In other words, all other nodes are descendants of the root. The sequence of arcs from a node to its descendants is a *path*.

The *depth* (or *level*) *of a node* is zero for a root node, and one more than the depth of its parent for a non-root node. If the depth of a node x_i is less than the depth of a node x_j, then x_i is *shallower* than x_j, and x_j is *deeper* than x_i. The *depth of a tree* is equal to the depth of its deepest node.

2.2   Building a Decision Tree

A *decision tree* is a kind of tree which, in the training phase is built from a training set. Then in the testing phase, the tree is used to classify patterns. We can build more than one decision tree from a given training set. The procedure that follows builds an arbitrary decision tree. Later, Section 2.5 develops a criterion for selecting the tree to be built.

This procedure builds the tree in a *breadth-first* manner: shallower nodes are generated before deeper nodes. We begin by generating the root node. It is then expanded to generate all first-level nodes. The first-level nodes are then expanded one by one to generate all second-level nodes, which are then expanded one by one to generate all third-level nodes, and so on.

The procedure uses two lists OPEN and CLOSED. At any time during the execution of the procedure, unexpanded nodes are stored in OPEN (they are still open to expansion), and nodes already expanded are stored in CLOSED (they are closed, or unavailable, for expansion). Each list has a front and a back: nodes are put at the back of the list, and they are removed from the front. In other words, both OPEN and CLOSED lists store nodes as queues do. In expanding a node, not only do we generate its children, but we also transfer it from OPEN to CLOSED.

Let V denote a training set. As mentioned in Section 1.1, let us assume that the training patterns have at least one attribute, each attribute has at least two possible values, the number of classes is $m \geq 1$, and each pattern belongs to exactly one class. The breadth-first procedure to build a decision tree from a given training set is as follows.

1. If $m = 1$ (that is, there is only one class), then create a single node x, label x with the name of the single class, and terminate the procedure. The decision tree consists of the single node x, with its class label.

2. Initialize lists OPEN and CLOSED to empty.

3. Initialize subscript i to empty. Create a root node x_i, and associate the training set V_i with it.

4. Put node x_i in OPEN.

5. If OPEN is empty, return from the procedure. The nodes in CLOSED constitute the decision tree built. The subscript and the label of each node in CLOSED, together with the labelled arc from the node's parent, delineate the decision tree (step 10.1 explains how the arcs are labelled).

6. Remove the frontmost node x_i from OPEN. Create a *candidate set of attributes* for x_i, where the set contains all those attributes that have not been examined at any node on the path from the root to x_i. Select an attribute A from the candidate set for examining at node x_i (the criterion developed in Section 2.5 can be used to select attributes; for now, it may be selected arbitrarily).

7. If subscript $i = $ empty (that is, x_i is the root), then put the following at the back of the list CLOSED: node x_i together with the attribute A being examined at it.

8. If $i \neq$ empty (that is, x_i is not the root), then put the following at the back of the list CLOSED: node x_i together with the attribute A being examined at it and the label of the arc from the parent-of-x_i to x_i.

9. Examine attribute A as follows. If A has n possible values $v_1, v_2, \ldots, v_n$, then expand node x_i to generate its n children $x_{i1}, x_{i2}, \ldots, x_{in}$.

10. For $j = 1, 2, \ldots, n$ do steps 10.1 to 10.8.

   10.1. Label the arc from node x_i to node x_{ij} with attribute value v_j.

   10.2. Associate with x_{ij} the set $V_{ij} \subseteq V_i$, such that the value of attribute A for the patterns in V_{ij} is v_j.

   10.3. If all the patterns in set V_{ij} belong to one class, then label node x_{ij} with the name of that class. Go to step 10.7.

   10.4. If V_{ij} is empty, then label node x_{ij} with '?' to indicate rejection, that is, failure to classify a given pattern. Go to step 10.7. (The question mark symbol, '?', has been used to indicate rejection, which means that a given pattern has not been classified. You can also replace it by a symbol of your choice.)

10.5. If the patterns in V_{ij} belong to more than one class and all the attributes have been examined on the path from the root to x_i, then label x_{ij} with '?' to indicate rejection. (One could say that ideally we should have more attributes available to put the patterns of V_{ij} into separate classes, but more attributes may not be available. An alternative to rejection is to label node x_{ij} with the probability of the occurrence of different classes in V_{ij}; more on this is given in Chapter 3.) Go to step 10.7.

10.6. Put the following at the back of the list OPEN: the node x_{ij} (it is a non-leaf node), and the label of the arc from node x_i to x_{ij}. Go to step 10.8.

10.7. Put the following at the back of the list CLOSED: node x_{ij}, a mark to indicate that x_{ij} is a leaf node, the label of x_{ij}, and the label of the arc from x_i to x_{ij}.

10.8. Continue.

11. Go to step 5.

On the path from the root to a leaf node, no attribute is examined more than once. It is, however, not necessary that every attribute be examined before we arrive at a leaf node, as can happen in steps 10.3 and 10.4 of this procedure. The depth of a leaf node will therefore be less than or equal to the number of attributes. Since, by definition, the depth of a tree is equal to the depth of its deepest node, the depth of the tree, too, will be less than or equal to the number of attributes.

As an example, let us build a decision tree using the professor–student training set given in Fig. 1.1. Of the eight patterns in the training set, patterns T2, T4, and T7 belong to the class P (for professor), and patterns T1, T3, T5, T6, and T8 belong to the class S (for student). We associate training set V with node x, which will become the root of the decision tree. To understand this better, it is advisable to write down the contents of the training set V next to node x given in Fig. 2.1.

The candidate set of attributes at node x consists of HABIT, EATS, and FOOTWEAR. Let us—for the time being, arbitrarily—select HABIT to be examined at node x (the criterion for selecting an attribute is developed in Section 2.5). Attribute HABIT has two possible values: gabby and quiet. We expand x to generate its two children x_1 and x_2. We label the arc from x to x_1 as gabby, and the arc from x to x_2 as quiet.

We have labelled the arc from x to x_1 as gabby, so we associate set $V_1 = \{T1, T2, T3, T5, T7\}$ with node x_1 (because from the set V, the patterns in V_1 have HABIT equal to gabby). Similarly, since we have labelled the arc from x to x_2 as quiet, we associate set $V_2 = \{T4, T6, T8\}$ with node x_2 (because from the set V, the patterns in V_2 have HABIT equal to quiet).

Since HABIT has already been examined at the parent of node x_1, the candidate set of attributes at x_1 consists of EATS and FOOTWEAR. Let us, arbitrarily, select EATS to be examined at x_1. Attribute EATS has three possible values: baked, fried, and roasted. We expand x_1 to generate its three children x_{11}, x_{12}, and x_{13}, and label the three arcs from x_1 to its children as baked, fried, and roasted, respectively. We associate set $V_{11} = \{T1, T3\}$ with node x_{11} (because from the set V_1, the patterns in V_{11} have EATS equal to baked). Since all the patterns in V_{11} belong to the class student, node x_{11} becomes a leaf node (see step 10.3 of the above procedure to build decision trees), and we label it with S (for student).

The reader can complete the decision tree for this example following the procedure mentioned above. Figure 2.2 contains a brief explanation for generating each node of the tree and can be referred to if there is any difficulty in completing the decision tree.

The decision tree built should finally look like the tree given in Fig. 2.3. The following minor changes have been made to this tree to make it easier to understand.

- The names of the nodes (x, x_1, x_2, ...) and the names of the sets (V, V_1, V_2, ...) associated with the nodes have not been shown to avoid cluttering the tree.
- At a non-leaf node, a question mark has been put after the name of the attribute being examined to indicate that the question of the class of the pattern has not yet been answered, and that we are looking at the possible values of that attribute.

Node	Set associated with the node	Attribute examined	Arc from node to
x (root)	{T1, T2, T3, T4, T5, T6, T7, T8}	HABIT = gabby	x_1
		= quiet	x_2
x_1	{T1, T2, T3, T5, T7}	EATS = baked	x_{11}
		= fried	x_{12}
		= roasted	x_{13}
x_2	{T4, T6, T8}	EATS = baked	x_{21}
		= fried	x_{22}
		= roasted	x_{23}
x_{11}	{T1, T3}	None, since node becomes a leaf node with label S	No other node
x_{12}	{T5, T7}	FOOTWEAR = clogs	x_{121}
		= sandals	x_{122}
x_{13}	{T2}	None, since node becomes a leaf node with label P	No other node
x_{21}	{T6}	None, since node becomes a leaf node with label S	No other node
x_{22}	{T4, T8}	FOOTWEAR = clogs	x_{221}
		= sandals	x_{222}
x_{23}	{ }	None, since node becomes a leaf node with label '?'	No other node
x_{121}	{T5}	None, since node becomes a leaf node with label S	No other node
x_{122}	{T7}	None, since node becomes a leaf node with label P	No other node
x_{221}	{T8}	None, since node becomes a leaf node with label S	No other node
x_{222}	{T4}	None, since node becomes a leaf node with label P	No other node

FIGURE 2.2   Generating nodes for a decision tree from the professor–student training set given in Fig. 1.1. The symbols '{ }' denote an empty set and '?' a rejection. A leaf node is labelled S for the class student and P for the class professor. The decision tree is shown in Fig. 2.3.

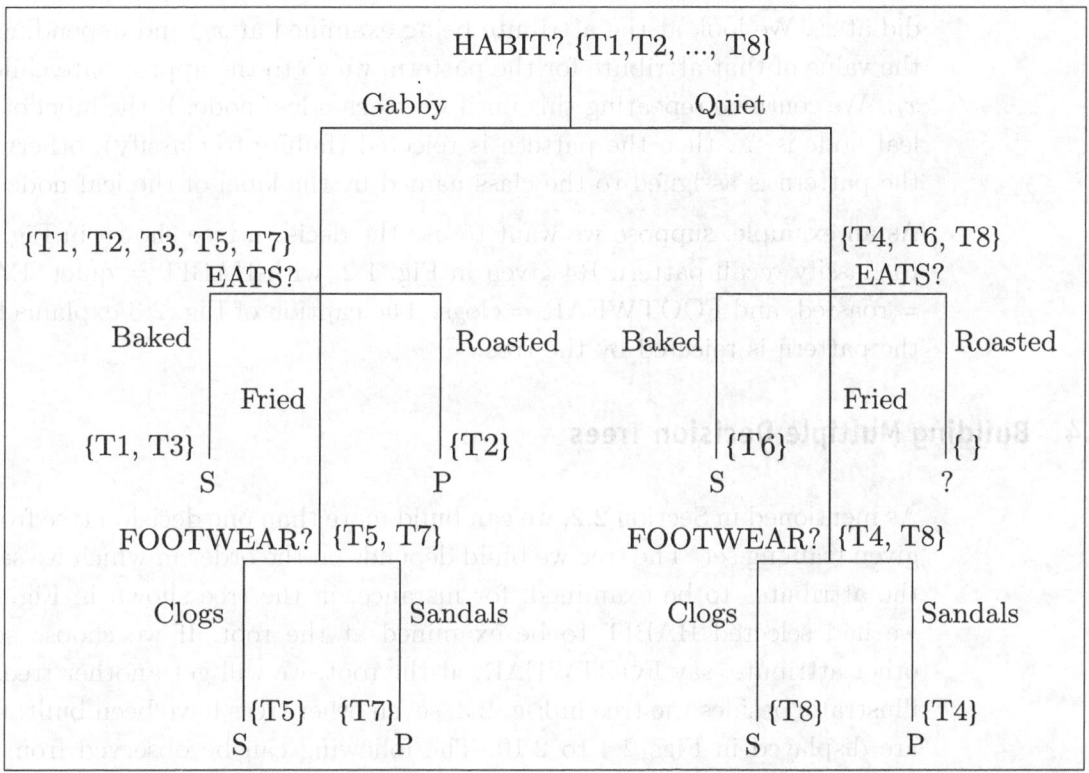

FIGURE 2.3 Decision tree built from the professor–student training set given in Fig. 1.1. Figure 2.2 shows how each node was obtained. At a leaf node, P denotes the class professor and S denotes the class student. To classify the recall pattern R4 of Fig. 1.2 (HABIT = quiet, EATS = roasted, and FOOTWEAR = clogs), we start from the root and examine the value of HABIT for the pattern; since the value is quiet, we arrive at the right child of the root. At the child, we examine the value of EATS; since it is roasted, we arrive at the leaf node labelled '?'. The pattern is rejected even without examining its attribute FOOTWEAR.

2.3 Classifying by Using Decision Trees

Suppose we are given a pattern (be it recall or training), which means we are given the value of its various attributes and we want to classify the pattern by using an available decision tree.

We start from the root of the tree. At any non-leaf node x_i, we look at the attribute being examined. Suppose the possible values of that attribute are $v_1, v_2, \ldots, v_n$, and the value of the attribute for the pattern is v_j. Then we go to x_{ij}, the jth child of x_i. If x_{ij} is a non-leaf node, we repeat what we did at x_i. We look at the attribute being examined at x_{ij} and depending on the value of that attribute for the pattern, we go to the appropriate child of x_{ij}. We continue repeating this until we reach a leaf node. If the label of the leaf node is '?', then the pattern is rejected (failure to classify); otherwise, the pattern is assigned to the class named by the label of the leaf node.

As an example, suppose we want to use the decision tree shown in Fig. 2.3 to classify recall pattern R4 given in Fig. 1.2, with HABIT = quiet, EATS = roasted, and FOOTWEAR = clogs. The caption of Fig. 2.3 explains how the pattern is rejected by the tree.

2.4 Building Multiple Decision Trees

As mentioned in Section 2.2, we can build more than one decision tree from a given training set. The tree we build depends on the order in which we select the attributes to be examined; for instance, in the tree shown in Fig. 2.3, we had selected HABIT to be examined at the root. If we choose some other attribute, say FOOTWEAR, at the root, we will get another tree. To illustrate, besides the tree in Fig. 2.3, seven other trees have been built: they are displayed in Figs 2.4 to 2.10. The following can be observed from the eight trees shown in Figs 2.3 to 2.10 if one studies each tree and reads its caption.

- It is not necessary that the same attribute be examined at sibling nodes: in Fig. 2.5, at level 1, one node examines EATS, while the other node examines FOOTWEAR. Similarly, the sibling nodes in Fig. 2.6 examine different attributes.
- The three decision trees in Figs 2.7 to 2.9 do not reject any pattern, but the remaining trees do.
- The two decision trees shown in Figs 2.7 and 2.9 are of depth two and the remaining trees are of depth three.

From the captions of the decision trees given in Figs 2.3 to 2.10, we can see that, as an example, when we classify recall pattern R4 given in Fig. 1.2 (HABIT = quiet, EATS = roasted, and FOOTWEAR = clogs) using the eight different trees, we do not obtain an identical classification— Figs 2.3 and 2.6 reject it; Figs 2.4, 2.5, 2.9, and 2.10 classify it in the class student; and Figs 2.7 and 2.8 classify it in the class professor. In general, a given pattern can be classified differently by different decision trees. The classifications of the four recall patterns given in Fig. 1.2 according to the

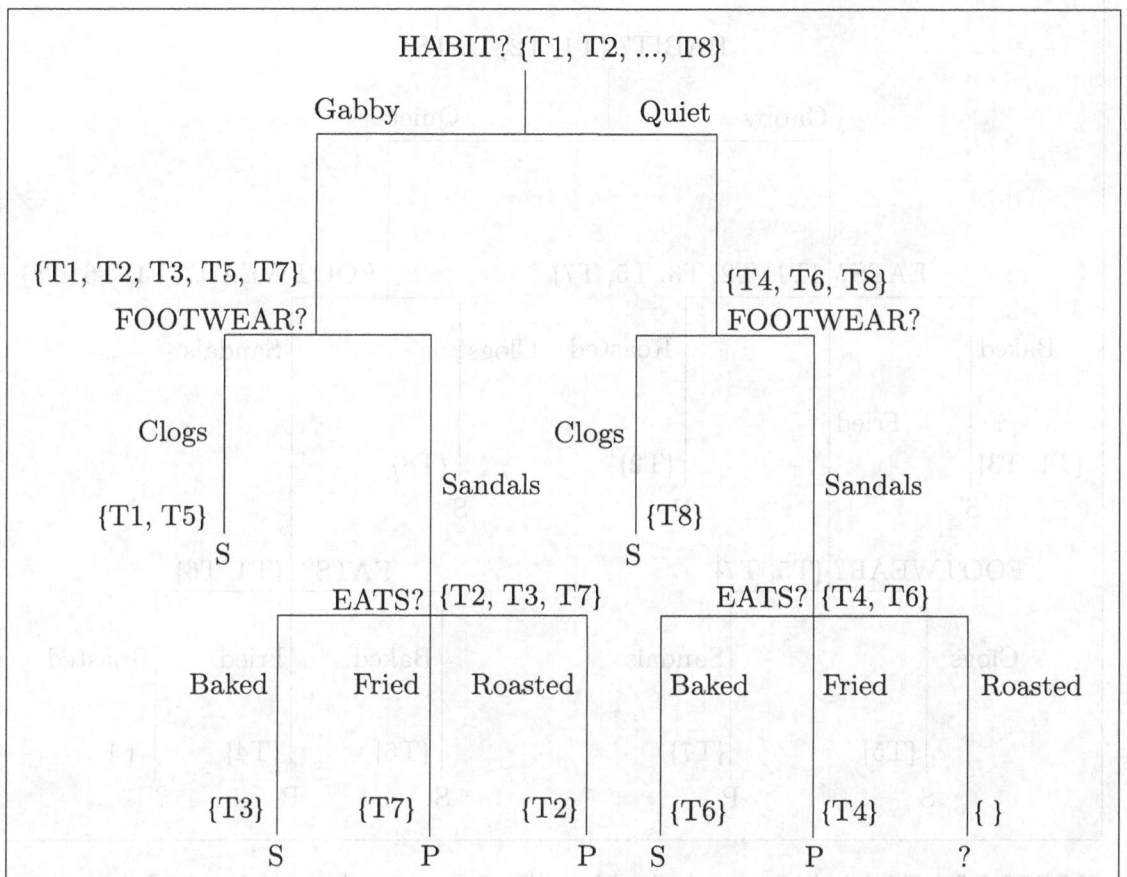

FIGURE 2.4 Second decision tree obtained from the professor–student training set given in Fig. 1.1. The recall pattern R4 given in Fig. 1.2 (HABIT = quiet, EATS = roasted, and FOOTWEAR = clogs) is classified by the above tree as student. This is different from the result given by the decision tree in Fig. 2.3, where the pattern is rejected.

eight trees are given in Fig. 2.11. It is useful to verify these results by working on them. It is also interesting to observe that recall pattern R1 has been classified as student by all the eight trees, but the classifications of patterns R2 to R4 vary.

It will not be surprising if different people are asked to classify intuitively the recall patterns given in Fig. 1.2, and they come out with different results. It can be assumed that they have intuitively built different decision trees.

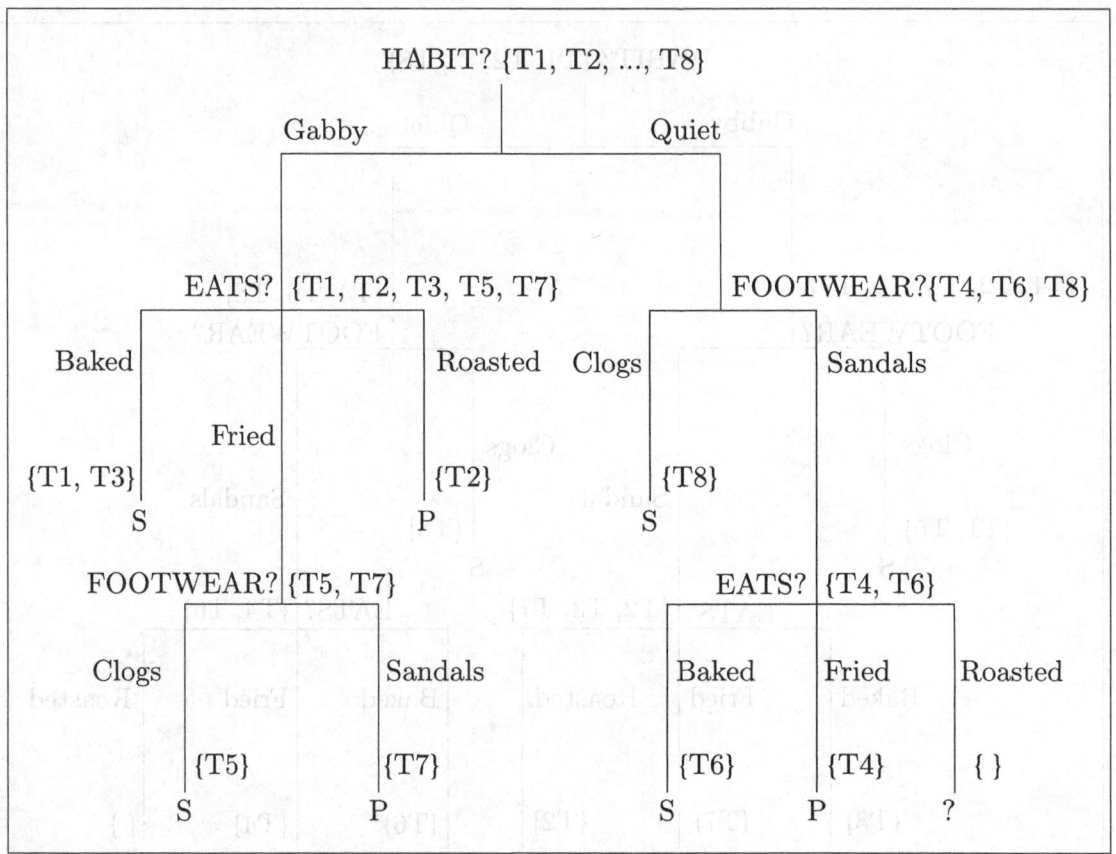

FIGURE 2.5   Third decision tree obtained from the professor–student training set. It is not necessary that the same attribute be examined at sibling nodes—in this figure, at depth 1, we examine EATS at one node, and FOOTWEAR at its sibling. The recall pattern R4 is classified by the above tree as student. This result is the same as given by the decision tree shown in Fig. 2.4, but different from the result given by the decision tree shown in Fig. 2.3, where the pattern is rejected.

2.5   Selecting the Decision Tree to be Built

While reading the professor–student training set example in Chapter 1, it might not have become immediately evident that attribute HABIT is redundant and that it need not be examined to classify a pattern. The decision trees shown in Figs 2.7 and 2.9, however, confirm it. These decision trees are of depth two because they examine only two attributes (EATS and FOOTWEAR). The other six decision trees (Figs 2.3 to 2.6, 2.8, and 2.10) are of depth three because they examine all the three attributes (HABIT, EATS, and FOOTWEAR). In general, shallow decision trees examine fewer attributes than deep trees. Since examining each attribute requires some computation, the computation required by shallow trees is expected to be lower than that required by deep trees. Accordingly, for a given classification problem, if given a choice, a shallow decision tree should be preferred over a deep decision tree. Selecting the tree in which fewer attributes are examined is another instance of obeying the minimum description length principle (mentioned in Section 1.5):

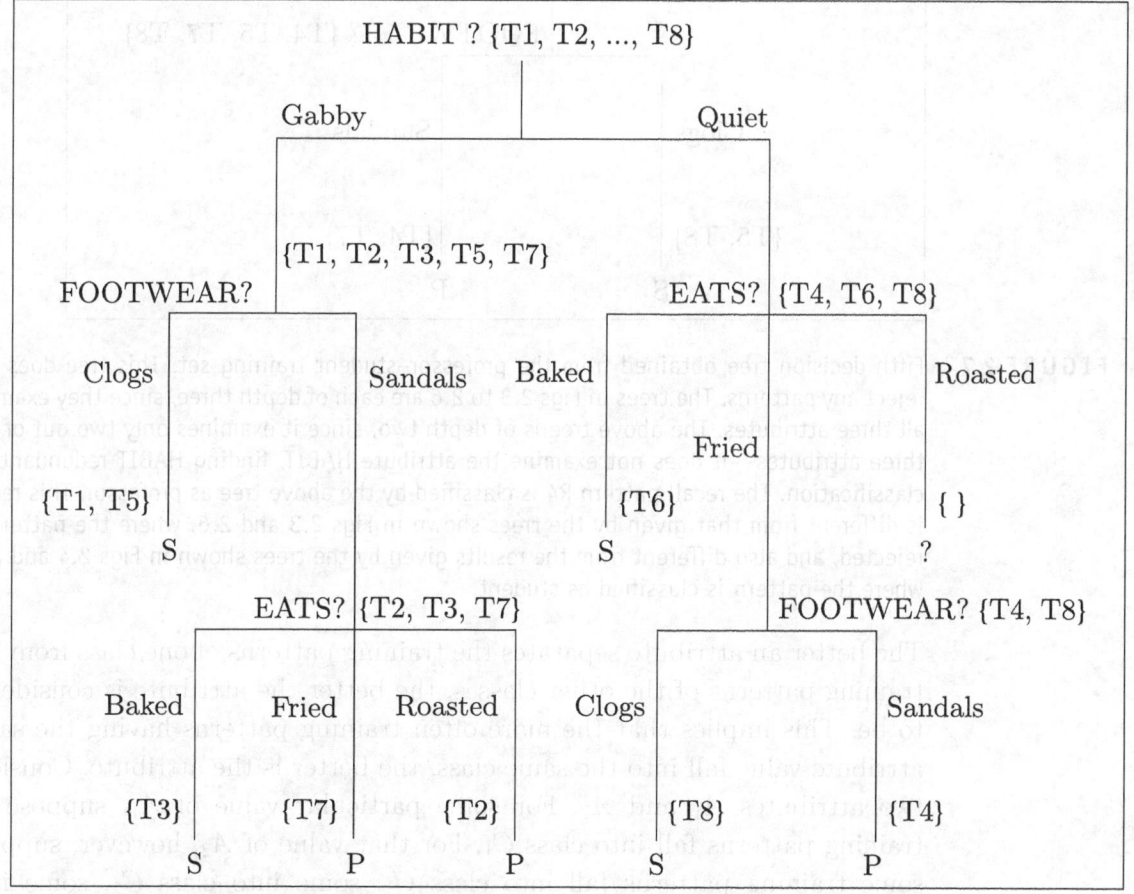

FIGURE 2.6   Fourth decision tree obtained from the professor–student training set. The recall pattern R4 is rejected by the above tree. This result is the same as that given by the decision tree shown in Fig. 2.3, but different from that given by the decision trees shown in Figs 2.4 and 2.5, where the pattern is classified as student.

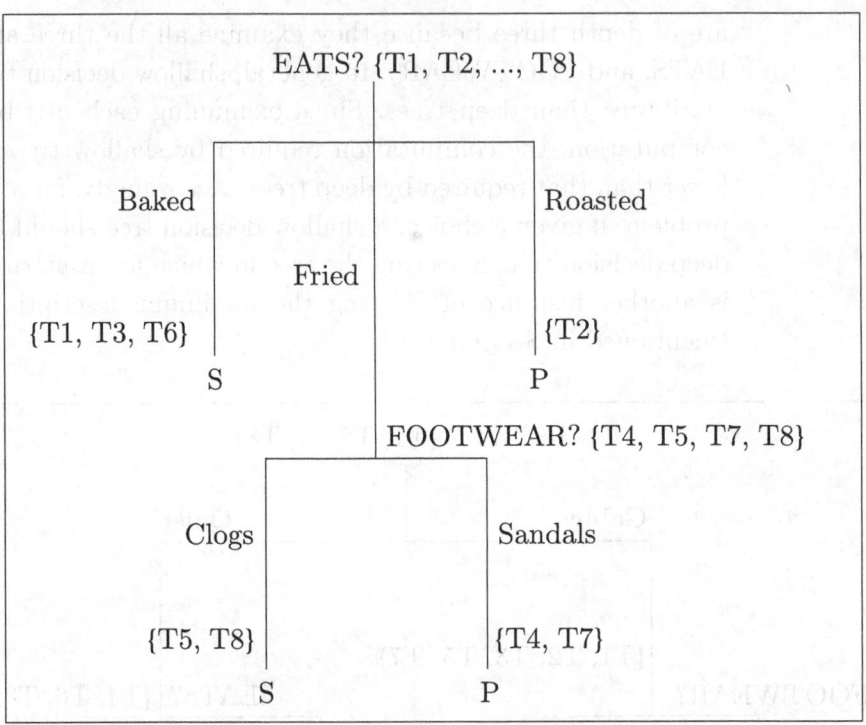

FIGURE 2.7   Fifth decision tree obtained from the professor–student training set. This tree does not reject any patterns. The trees in Figs 2.3 to 2.6 are each of depth three, since they examine all three attributes. The above tree is of depth two, since it examines only two out of the three attributes—it does not examine the attribute HABIT, finding HABIT redundant for classification. The recall pattern R4 is classified by the above tree as professor. This result is different from that given by the trees shown in Figs 2.3 and 2.6, where the pattern is rejected, and also different from the results given by the trees shown in Figs 2.4 and 2.5, where the pattern is classified as student.

The better an attribute separates the training patterns of one class from the training patterns of the other classes, the better the attribute is considered to be. This implies that the more often training patterns having the same attribute value fall into the same class, the better is the attribute. Consider two attributes A_1 and A_2. For some particular value of A_1, suppose all training patterns fall into class C_1. For that value of A_2, however, suppose some training patterns fall into class C_1, some into class C_2, some into class C_3, and so on. Then for this attribute value, A_1 separates the training patterns better than A_2. If A_1 does so for other values as well, then attribute A_1 is better than attribute A_2. In general, the more unevenly an attribute distributes training patterns over different classes, the better the attribute is.

Whenever we examine an attribute at some node of a decision tree, we gain some information about the class of a given pattern. The information gained is a function of the unevenness with which the attribute distributes the training patterns over the different classes. Better attributes provide a higher ratio of information gain compared with other attributes. From the candidate set of attributes for a given non-leaf node in a decision tree, we can choose to examine the attribute that maximizes this ratio. A decision

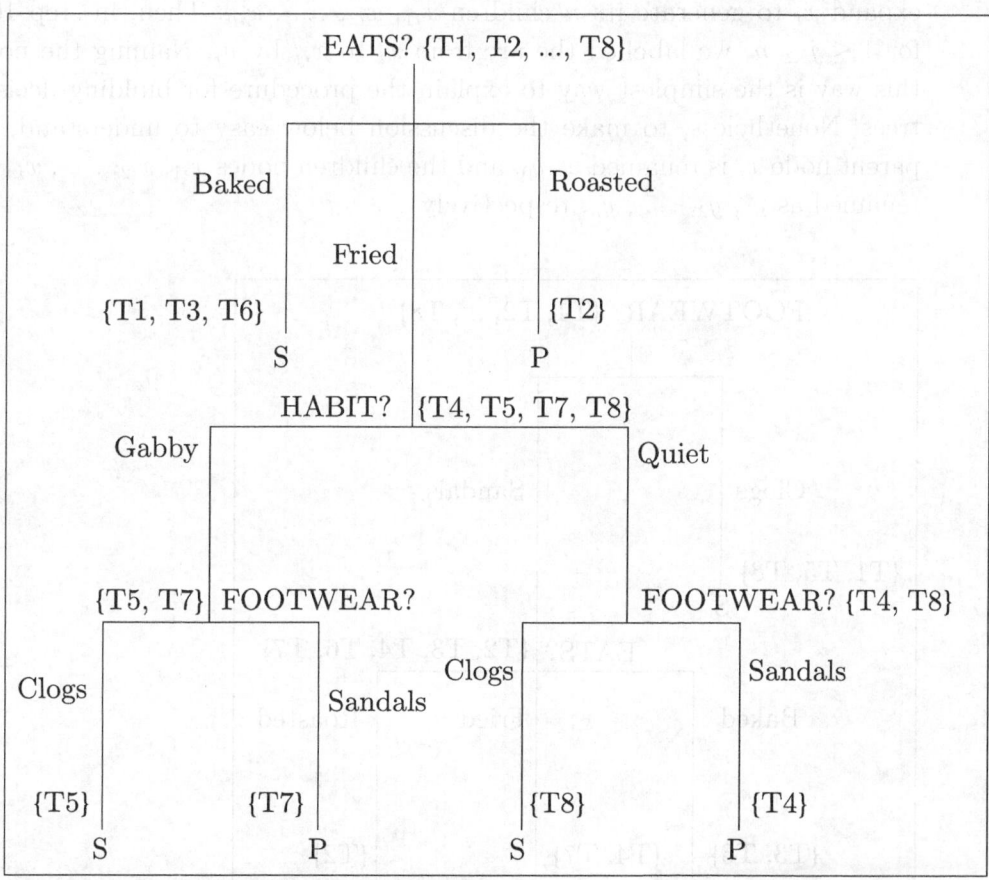

FIGURE 2.8 Sixth decision tree obtained from the professor–student training set. Like the tree shown in Fig. 2.7, the above tree does not reject any patterns. However, the tree shown in Fig. 2.7 has depth two (it examines only two out of the three attributes), whereas the above tree has depth three (it examines all three attributes). The recall pattern R4 is classified by the above tree as professor. This result is the same as that given by the decision tree shown in Fig. 2.7, but different from that given by the tree shown in Figs 2.3 and 2.6, where the pattern is rejected, and also different from the results given by the trees of Figs 2.4 and 2.5, where the pattern is classified as student.

tree built by maximizing this ratio for each of the tree's non-leaf nodes is typically shallower than a tree built otherwise. The description below develops a mathematical equation to evaluate this ratio for a given attribute A being examined at some node.

In step 9 of the procedure to build decision trees (Section 2.2), we examined attribute A at a node x_i. If A has n possible values $v_1, v_2, \ldots, v_n$, we expand x_i to generate its n children $x_{i1}, x_{i2}, \ldots, x_{in}$. Then, in step 10.1, for $1 \leq j \leq n$, we labelled the arc from x_i to x_{ij} by v_j. Naming the nodes this way is the simplest way to explain the procedure for building decision trees. Nonetheless, to make the discussion below easy to understand, the parent node x_i is renamed as y_0, and the children nodes $x_{i1}, x_{i2}, \ldots, x_{in}$ are renamed as $y_1, y_2, \ldots, y_n$, respectively.

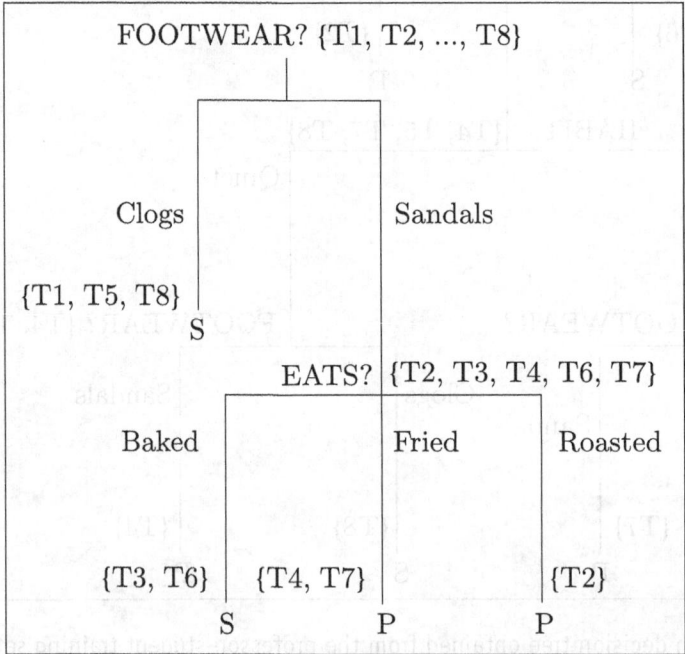

FIGURE 2.9   Seventh decision tree obtained from the professor–student training set. Like the tree shown in Fig. 2.7, the above tree is also of depth two (it examines only two out of the three attributes) and it does not reject any patterns. The recall pattern R4 is classified by the above tree as student. This result is the same as given by the decision trees shown in Figs 2.4 and 2.5, but different from that given by the decision trees shown in Figs 2.3 and 2.6, where the pattern is rejected, and also different from that given by the trees shown in Figs 2.7 and 2.8, where the pattern is classified as professor.

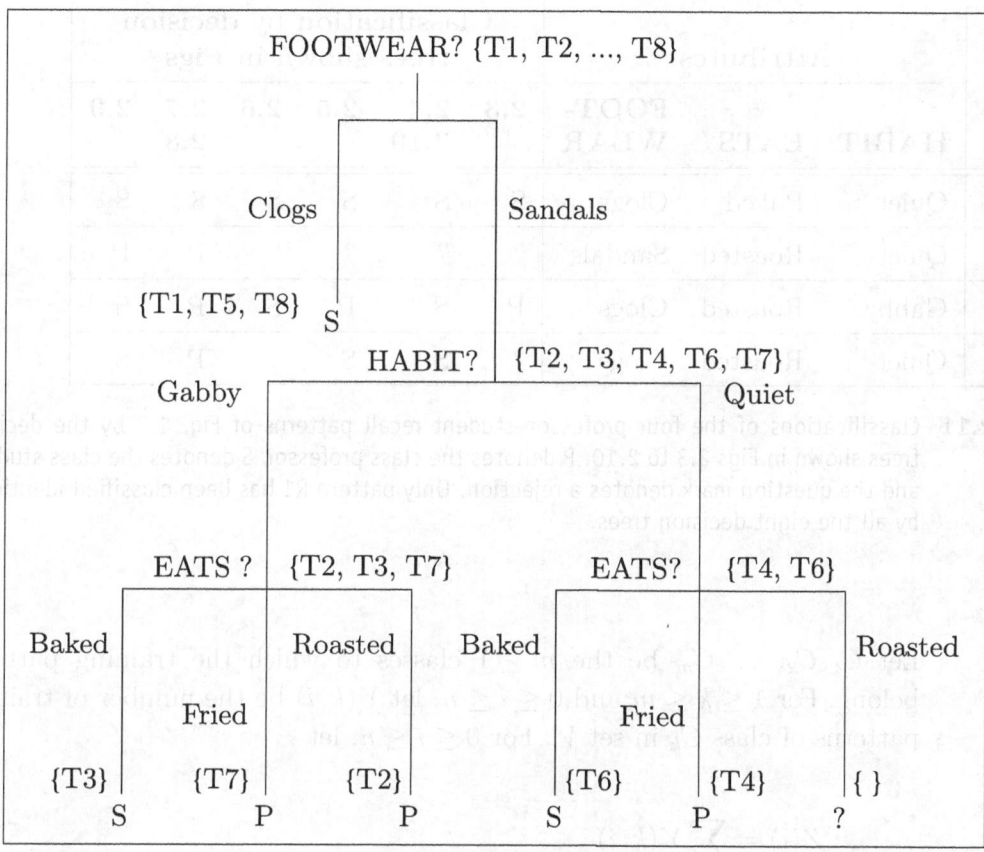

FIGURE 2.10 Eighth decision tree obtained from the professor–student training set. The recall pattern R4 is classified by the above tree as student. This result is the same as that given by the decision trees shown in Figs 2.4, 2.5, and 2.9. It is, however, different from the result given by the decision trees shown in Figs 2.3 and 2.6, where the pattern is rejected, and also different from the results given by the trees shown in Figs 2.7 and 2.8, where the pattern is classified as professor.

According to the procedure to build decision trees, a subset of the training set is associated with each node: for $0 \leq i \leq n$, a set V_i is associated with node y_i. At node y_0, we examine attribute A on the patterns of set V_0. Then, for $1 \leq j \leq n$, we have $V_j \subseteq V_0$, such that the value of attribute A for the patterns in V_j is v_j, where v_j is the label of the arc from node y_0 to node y_j. Set V_0 is thus the union of the sets V_1 to V_n. We can alternatively say that attribute A *splits* set V_0 into sets V_1 to V_n.

Name of recall pattern	Attributes			Classification by decision trees shown in Figs					
	HABIT	**EATS**	**FOOT-WEAR**	2.3 2.10	2.4	2.5	2.6	2.7 2.8	2.9
R1	Quiet	Baked	Clogs	S	S	S	S	S	S
R2	Quiet	Roasted	Sandals	?	?	?	?	P	P
R3	Gabby	Roasted	Clogs	P	S	P	S	P	S
R4	Quiet	Roasted	Clogs	?	S	S	?	P	S

FIGURE 2.11 Classifications of the four professor–student recall patterns of Fig. 1.2 by the decision trees shown in Figs 2.3 to 2.10. P denotes the class professor, S denotes the class student, and the question mark denotes a rejection. Only pattern R1 has been classified identically by all the eight decision trees.

Let $C_1, C_2, \ldots, C_m$ be the $m \geq 1$ classes to which the training patterns belong. For $1 \leq k \leq m$ and $0 \leq i \leq n$, let $Y(k, i)$ be the number of training patterns of class C_k in set V_i. For $0 \leq i \leq n$, let

$$Z(i) = \sum_{k=1}^{m} Y(k, i)$$

which means that $Z(i)$ is equal to the number of patterns in the set V_i, that is, $Z(i)$ is equal to the cardinality of V_i.

We can now develop the equation for the ratio of information gain provided by attribute A when it is examined at node y_0. For $1 \leq k \leq m$, the probability that a pattern in V_0 belongs to class C_k is

$$\frac{Y(k, 0)}{Z(0)}$$

The lower the above probability, the more the uncertainty in classifying a pattern in set V_0 into class C_k. The more the uncertainty, the more the information required to classify. Thus, the lower the above probability, the more the need for such information. The information required to classify a pattern in V_0 into class C_k, for $1 \leq k \leq m$, is then expressed as

$$- \log \frac{Y(k,0)}{Z(0)}$$

By convention, the information is measured in bits, and so the base of the above log is 2, as will be the case for all logs throughout the current discussion. (The above equation is rooted in information theory; readers who are not familiar with this area, can look up a book on information theory; A discourse on information theory will distract from the current discussion.)

The average information required to classify a pattern in set V_0 into one of the m classes is then expressed as

$$I(V_0) = - \sum_{k=1}^{m} \frac{Y(k,0)}{Z(0)} \log \frac{Y(k,0)}{Z(0)} \tag{2.1}$$

$I(V_0)$ is also known as the *entropy of the set V_0*. An example to calculate the entropy is given in Fig. 2.12. Proceeding similarly for $1 \leq j \leq n$, the average information required to classify a pattern in set V_j into one of the m classes

$$I(V_j) = - \sum_{k=1}^{m} \frac{Y(k,j)}{Z(j)} \log \frac{Y(k,j)}{Z(j)}$$

The average information required to classify a pattern belonging to set V_0 after it has been split by attribute A into sets V_1 to V_n is

$$I_A(V_0) = \sum_{j=1}^{n} \frac{Z(j)}{Z(0)} I(V_j)$$

Combining the last two equations, we get

$$I_A(V_0) = - \sum_{j=1}^{n} \frac{Z(j)}{Z(0)} \sum_{k=1}^{m} \frac{Y(k,j)}{Z(j)} \log \frac{Y(k,j)}{Z(j)} \tag{2.2}$$

To evaluate the entropy $I(V_0)$ of the professor–student training set of Fig. 1.1. The classes are C_1 (professor) and C_2 (student). Therefore, $m = 2$. Three patterns belong to C_1 and five to C_2. Therefore,

$$\text{The training set } V_0 = \{\text{T1, T2, } \ldots \text{, T8}\}$$
$$Y(1,0) = 3 \text{ (number of patterns in } V_0 \text{ of class } C_1)$$
$$Y(2,0) = 5 \text{ (number of patterns in } V_0 \text{ of class } C_2)$$
$$Z(0) = 8 \text{ (number of patterns in } V_0)$$

From Eqn (2.1) in Section 2.5, the average information required to classify a pattern in V_0 into one of the m classes is

$$I(V_0) = -\sum_{k=1}^{m} \frac{Y(k,0)}{Z(0)} \log \frac{Y(k,0)}{Z(0)}$$

$$= -\sum_{k=1}^{2} \frac{Y(k,0)}{Z(0)} \log \frac{Y(k,0)}{Z(0)}$$

$$= -\frac{3}{8} \log \frac{3}{8} - \frac{5}{8} \log \frac{5}{8}$$

$$= 0.375 \times 1.415 + 0.625 \times 0.6780$$

$$= 0.5306 + 0.4238$$

$$= 0.9544$$

FIGURE 2.12 Evaluating entropy $I(V_0)$, where V_0 is the professor–student training set of Fig. 1.1.

$I_A(V_0)$ is also known as the *entropy of the attribute A* for set V_0. The *gain in information* caused by attribute A splitting set V_0 into sets V_1 to V_n is then defined as

$$g_A(V_0) = I(V_0) - I_A(V_0) \tag{2.3}$$

We want to maximize this gain in information. For a given non-leaf node, we calculate the gain in information for each attribute in the node's candidate set of attributes. The attribute that has the highest gain can then be selected for examination at the node. This criterion of selection can be used in step 6 of the procedure to build decision trees, described in Section 2.2.

There are, however, occasions when maximizing this gain can be misleading. The professor–student training patterns given in Fig. 1.1 have been arbitrarily given the names T1 to T8. Suppose we were to consider NAME to be an attribute. At the root, NAME will split set V_0 into eight sets V_1 to V_8, where each V_j, for $1 \leq j \leq 8$, will contain one pattern. The patterns in sets V_2, V_4, and V_7 will belong to class C_1 (professor), and the patterns in the remaining sets will belong to class C_2 (student). The entropy of the attribute NAME will be zero, and consequently it will maximize the gain in information. Nonetheless, the attribute NAME will be useless in classifying recall patterns—since each pattern has a unique NAME value, no recall pattern will have a NAME value that is identical to the NAME value of a training pattern. A decision tree built by examining NAME at the root will reject all recall patterns. For an attribute to be useful, patterns should share attribute values. By maximizing information gain, we bias our selection in favour of the attribute that splits set V_0 into more non-empty subsets and against the attribute that splits V_0 into fewer non-empty subsets.

To rectify this bias, we need to also consider the attribute's *split information* (how the attribute splits set V_0). The probability that there are $Z(j)$ patterns of set V_j among the $Z(0)$ patterns of set V_0 is

$$\frac{Z(j)}{Z(0)}$$

The information needed to extract set V_j from V_0 is

$$-\log \frac{Z(j)}{Z(0)}$$

So, the average information needed by attribute A to split set V_0 into sets V_1 to V_n is

$$S_A(V_0) = -\sum_{j=1}^{n} \frac{Z(j)}{Z(0)} \log \frac{Z(j)}{Z(0)} \tag{2.4}$$

The *ratio of information gain* of attribute A for set V_0 is then defined as

$$G_A(V_0) = \frac{g_A(V_0)}{S_A(V_0)}$$

Then, from Eqn (2.3), we get the ratio of information gain

$$G_A(V_0) = \frac{I(V_0) - I_A(V_0)}{S_A(V_0)} \tag{2.5}$$

We want to maximize this ratio of information gain. For a given non-leaf node, we calculate this ratio for each attribute in the node's candidate set of attributes. The attribute that has the highest ratio is then selected for examination at the node. This criterion of selection can be used in step 6 of the procedure to build decision trees, described in Section 2.2.

When we build a decision tree, as the nodes become deeper and deeper, the training set gets split further and further. Thus, the deeper the non-leaf node, the fewer the training patterns associated with it, and less is the theoretical justification required in using the ratio of information gain to decide on the attribute to be examined at that node. In practice, however, the ratio is used at deep nodes as well. As an example, we will use the ratio to build the decision tree for the professor–student training set given in Fig. 1.1.

1. The candidate set of attributes that can be examined at the root consists of HABIT, EATS, and FOOTWEAR. From Figs 2.13 to 2.15, we see that, at the root, $G_{\text{HABIT}}(V_0) = 0.0047$, $G_{\text{EATS}}(V_0) = 0.3233$, and $G_{\text{FOOTWEAR}}(V_0) = 0.3640$. Since $G_{\text{FOOTWEAR}}(V_0)$ is the largest of the three ratios of information gain, we select FOOTWEAR to be examined at the root. Accordingly, the decision tree we will build will look either like the tree in Fig. 2.9, or like the tree in Fig. 2.10.

2. Examining FOOTWEAR at the root, we notice, from Figs 2.9 and 2.10, that the set associated with the root's left child is {T1, T5, T8}. Every pattern in the set belongs to the class student; hence the left child becomes a leaf node labelled S for student. The set associated with the root's right child is then {T2, T3, T4, T6, T7}, out of which two patterns T3 and T6 belong to the class student, and the remaining three patterns belong to the class professor.

3. The candidate set of attributes that can be examined at the root's right child consists of HABIT and EATS, attribute FOOTWEAR having been already examined at the root. From Figs 2.16 and 2.17 we see that, at the right child of the root, $G_{\text{EATS}}(V_0) = 0.6377$ and $G_{\text{HABIT}}(V_0) = 0.0208$. Since $G_{\text{EATS}}(V_0)$ is the larger of the two ratios

of information gain, we select EATS to be examined at the right child of the root. The nodes generated by expanding the root's right child are all leaf nodes, and we obtain the decision tree shown in Fig. 2.9.

As the above example shows, using the criterion of maximizing ratio of information gain to build a decision tree, may require considerable computation. This criterion can still be recommended because such computation is required to be carried out only once to build the decision tree. Such a tree, however, is expected to save a lot more computation when the tree is later repeatedly used to classify recall patterns. The learning program mentioned in Section 1.1 is the one that will build the appropriate decision tree, for a given training set.

2.6 Obtaining Prules From Decision Trees

The learning program is often extended to obtain prules from the decision tree built. Patterns are then classified using these prules. Each path from the root to a leaf node provides a prule. The consequent of the prule contains the label of the leaf node. The antecedent of the prule is the conjunction of the attribute values required to reach that leaf node. In Section 2.5, the decision tree shown in Fig. 2.9 was built by maximizing the ratio of information gain. The following four prules constitute the knowledge base obtained from this tree.

If FOOTWEAR = clogs,
then the pattern belongs to the class student.

If FOOTWEAR = sandals,
and EATS = baked,
then the pattern belongs to the class student.

If FOOTWEAR = sandals,
and EATS = fried,
then the pattern belongs to the class professor.

If FOOTWEAR = sandals,
and EATS = roasted,
then the pattern belongs to the class professor.

To evaluate $G_{\text{HABIT}}(V_0)$ at the root of a decision tree for the professor–student training set of Fig. 1.1, the following procedure is used. The root is named y_0. The classes are C_1 (professor) and C_2 (student). Therefore, $m = 2$. Since HABIT has two possible values (gabby and quiet), root y_0 has two children y_1 and y_2. Therefore, $n = 2$. To understand the following material easily, one can look at any one of the decision trees shown in Figs 2.3 to 2.6, since each of them examines HABIT at the root.

$$V_0 = \{\text{T1, T2, } \ldots \text{, T8}\} \text{ at node } y_0$$

$$Y(1,0) = 3 \text{ (number of patterns in } V_0 \text{ of class } C_1)$$

$$Y(2,0) = 5 \text{ (number of patterns in } V_0 \text{ of class } C_2)$$

$$Z(0) = 8 \text{ (number of patterns in } V_0)$$

$$V_1 = \{\text{T1, T2, T3, T5, T7}\} \text{ at node } y_1, \text{where HABIT} = \text{gabby}$$

$$Y(1,1) = 2 \text{ (number of patterns in } V_1 \text{ of class } C_1)$$

$$Y(2,1) = 3 \text{ (number of patterns in } V_1 \text{ of class } C_2)$$

$$Z(1) = 5 \text{ (number of patterns in } V_1)$$

$$V_2 = \{\text{T4, T6, T8}\} \text{ at node } y_2, \text{where HABIT} = \text{quiet}$$

$$Y(1,2) = 1 \text{ (number of patterns in } V_2 \text{ of class } C_1)$$

$$Y(2,2) = 2 \text{ (number of patterns in } V_2 \text{ of class } C_2)$$

$$Z(2) = 3 \text{ (number of patterns in } V_2)$$

$$I(V_0) = 0.9544 \text{ (from Fig. 2.12)}$$

$$I_{\text{HABIT}}(V_0) = -\sum_{j=1}^{n} \frac{Z(j)}{Z(0)} \sum_{k=1}^{m} \frac{Y(k,j)}{Z(j)} \log \frac{Y(k,j)}{Z(j)} = 0.9499 \text{ [from Eqn (2.2)]}$$

$$S_{\text{HABIT}}(V_0) = -\sum_{j=1}^{n} \frac{Z(j)}{Z(0)} \log \frac{Z(j)}{Z(0)} = 0.9544 \text{ [from Eqn (2.4)]}$$

$$G_{\text{HABIT}}(V_0) = \frac{I(V_0) - I_{\text{HABIT}}(V_0)}{S_{\text{HABIT}}(V_0)} = 0.0047 \text{ [from Eqn (2.5)]}$$

FIGURE 2.13 Evaluating $G_{\text{HABIT}}(V_0)$, the ratio of information gain of HABIT, at the root of a decision tree being built from the professor–student training set shown in Fig. 1.1.

To evaluate $G_{\text{EATS}}(V_0)$ at the root of a decision tree for the professor–student training set, the following procedure is used. The root is named y_0. The classes are C_1 (professor) and C_2 (student). Therefore, $m = 2$. Since EATS has three possible values (baked, fried, and roasted), root y_0 has three children y_1, y_2, and y_3. Therefore, $n = 3$. To understand the following material easily, one can look at either of the decision trees shown in Figs 2.7 or 2.8, since each of them examines EATS at the root.

$$V_0 = \{\text{T1, T2, } \ldots \text{, T8}\} \text{ at node } y_0$$
$$Y(1,0) = 3 \text{ (number of patterns in } V_0 \text{ of class } C_1)$$
$$Y(2,0) = 5 \text{ (number of patterns in } V_0 \text{ of class } C_2)$$
$$Z(0) = 8 \text{ (number of patterns in } V_0)$$
$$V_1 = \{\text{T1, T3, T6}\} \text{ at node } y_1, \text{ where EATS = baked}$$
$$Y(1,1) = 0 \text{ (number of patterns in } V_1 \text{ of class } C_1)$$
$$Y(2,1) = 3 \text{ (number of patterns in } V_1 \text{ of class } C_2)$$
$$Z(1) = 3 \text{ (number of patterns in } V_1)$$
$$V_2 = \{\text{T4, T5, T7, T8}\} \text{ at node } y_2, \text{ where EATS = fried}$$
$$Y(1,2) = 2 \text{ (number of patterns in } V_2 \text{ of class } C_1)$$
$$Y(2,2) = 2 \text{ (number of patterns in } V_2 \text{ of class } C_2)$$
$$Z(2) = 4 \text{ (number of patterns in } V_2)$$
$$V_3 = \{\text{T2}\} \text{ at node } y_2, \text{ where EATS = roasted}$$
$$Y(1,3) = 1 \text{ (number of patterns in } V_3 \text{ of class } C_1)$$
$$Y(2,3) = 0 \text{ (number of patterns in } V_3 \text{ of class } C_2)$$
$$Z(3) = 1 \text{ (number of patterns in } V_3)$$

$$I(V_0) = 0.9544 \text{ (from Fig. 2.12)}$$

$$I_{\text{EATS}}(V_0) = -\sum_{j=1}^{n} \frac{Z(j)}{Z(0)} \sum_{k=1}^{m} \frac{Y(k,j)}{Z(j)} \log \frac{Y(k,j)}{Z(j)} = 0.5 \text{ [from Eqn (2.2)]}$$

$$S_{\text{EATS}}(V_0) = -\sum_{j=1}^{n} \frac{Z(j)}{Z(0)} \log \frac{Z(j)}{Z(0)} = 1.4055 \text{ [from Eqn (2.4)]}$$

$$G_{\text{EATS}}(V_0) = \frac{I(V_0) - I_{\text{EATS}}(V_0)}{S_{\text{EATS}}(V_0)} = 0.3233 \text{ [from Eqn (2.5)]}$$

FIGURE 2.14 Evaluating $G_{\text{EATS}}(V_0)$, the ratio of information gain of EATS, at the root of a decision tree being built from the professor–student training set.

To evaluate $G_{\text{FOOTWEAR}}(V_0)$ at the root of a decision tree for the professor–student training set, the following procedure is used. The root is named y_0. The classes are C_1 (professor) and C_2 (student). Therefore, $m = 2$. Since FOOTWEAR has two possible values (clogs and sandals), root y_0 has two children y_1 and y_2. Therefore, $n = 2$. To understand the following material easily, one can look at either of the decision trees shown in Fig. 2.9 or 2.10, since each of them examines FOOTWEAR at the root.

$$V_0 = \{\text{T1, T2, } \ldots \text{, T8}\} \text{ at node } y_0$$

$Y(1,0) = 3$ (number of patterns in V_0 of class C_1)

$Y(2,0) = 5$ (number of patterns in V_0 of class C_2)

$Z(0) = 8$ (number of patterns in V_0)

$V_1 = \{\text{T1, T5, T8}\}$ at node y_1, where FOOTWEAR = clogs

$Y(1,1) = 0$ (number of patterns in V_1 of class C_1)

$Y(2,1) = 3$ (number of patterns in V_1 of class C_2)

$Z(1) = 3$ (number of patterns in V_1)

$V_2 = \{\text{T2, T3, T4, T6, T7}\}$ at node y_2, where FOOTWEAR = sandals

$Y(1,2) = 3$ (number of patterns in V_2 of class C_1)

$Y(2,2) = 2$ (number of patterns in V_2 of class C_2)

$Z(2) = 5$ (number of patterns in V_2)

$$I(V_0) = 0.9544 \quad \text{(from Fig. 2.12)}$$

$$I_{\text{FOOTWEAR}}(V_0) = -\sum_{j=1}^{n} \frac{Z(j)}{Z(0)} \sum_{k=1}^{m} \frac{Y(k,j)}{Z(j)} \log \frac{Y(k,j)}{Z(j)} = 0.6066 \quad [\text{from Eqn (2.2)}]$$

$$S_{\text{FOOTWEAR}}(V_0) = -\sum_{j=1}^{n} \frac{Z(j)}{Z(0)} \log \frac{Z(j)}{Z(0)} = 0.9544 \quad [\text{from Eqn (2.4)}]$$

$$G_{\text{FOOTWEAR}}(V_0) = \frac{I(V_0) - I_{\text{FOOTWEAR}}(V_0)}{S_{\text{FOOTWEAR}}(V_0)} = 0.3640 \quad [\text{from Eqn (2.5)}]$$

FIGURE 2.15 Evaluating $G_{\text{FOOTWEAR}}(V_0)$, the ratio of information gain of FOOTWEAR, at the root of a decision tree being built from the professor–student training set.

To evaluate $G_{\mathrm{EATS}}(V_0)$ at the right child of the root of a decision tree for the professor–student training set, the following procedure is used. FOOTWEAR has already been examined at the root (see Fig. 2.9). The right child of the root is named y_0. The classes are C_1 (professor) and C_2 (student). Therefore, $m = 2$. Since EATS has three possible values (baked, fried, and roasted), node y_0 has three children y_1, y_2, and y_3. Therefore, $n = 3$.

$$V_0 = \{\text{T2, T3, T4, T6, T7}\} \text{ at root } y_0$$
$$Y(1,0) = 3 \text{ (number of patterns in } V_0 \text{ of class } C_1)$$
$$Y(2,0) = 2 \text{ (number of patterns in } V_0 \text{ of class } C_2)$$
$$Z(0) = 5 \text{ (number of patterns in } V_0)$$
$$V_1 = \{\text{T3, T6}\} \text{ at node } y_1, \text{where EATS} = \text{baked}$$
$$Y(1,1) = 0 \text{ (number of patterns in } V_1 \text{ of class } C_1)$$
$$Y(2,1) = 2 \text{ (number of patterns in } V_1 \text{ of class } C_2)$$
$$Z(1) = 2 \text{ (number of patterns in } V_1)$$
$$V_2 = \{\text{T4, T7}\} \text{ at node } y_2, \text{where EATS} = \text{fried}$$
$$Y(1,2) = 2 \text{ (number of patterns in } V_2 \text{ of class } C_1)$$
$$Y(2,2) = 0 \text{ (number of patterns in } V_2 \text{ of class } C_2)$$
$$Z(2) = 2 \text{ (number of patterns in } V_2)$$
$$V_3 = \{\text{T2}\} \text{ at node } y_2, \text{where EATS} = \text{roasted}$$
$$Y(1,3) = 1 \text{ (number of patterns in } V_3 \text{ of class } C_1)$$
$$Y(2,3) = 0 \text{ (number of patterns in } V_3 \text{ of class } C_2)$$
$$Z(3) = 1 \text{ (number of patterns in } V_3)$$

$$I(V_0) = -\sum_{k=1}^{m} \frac{Y(k,0)}{Z(0)} \log \frac{Y(k,0)}{Z(0)} = 0.9710 \quad [\text{from Eqn (2.1)}]$$

$$I_{\mathrm{EATS}}(V_0) = -\sum_{j=1}^{n} \frac{Z(j)}{Z(0)} \sum_{k=1}^{m} \frac{Y(k,j)}{Z(j)} \log \frac{Y(k,j)}{Z(j)} = 0 \quad [\text{from Eqn (2.2)}]$$

$$S_{\mathrm{EATS}}(V_0) = -\sum_{j=1}^{n} \frac{Z(j)}{Z(0)} \log \frac{Z(j)}{Z(0)} = 1.5218 \quad [\text{from Eqn (2.4)}]$$

$$G_{\mathrm{EATS}}(V_0) = \frac{I(V_0) - I_{\mathrm{EATS}}(V_0)}{S_{\mathrm{EATS}}(V_0)} = 0.6377 \quad [\text{from Eqn (2.5)}]$$

FIGURE 2.16 Evaluating $G_{\mathrm{EATS}}(V_0)$, the ratio of information gain of EATS, at the right child of the root of a decision tree being built from the professor–student training set, FOOTWEAR having already been examined at the root, as in Fig. 2.9.

To evaluate $G_{\text{HABIT}}(V_0)$ at the right child of the root of a decision tree for the professor–student training set, the following procedure is used. FOOTWEAR has already been examined at the root (see Fig. 2.10). The right child of the root is named y_0. The classes are C_1 (professor) and C_2 (student). Therefore, $m = 2$. Since HABIT has two possible values (gabby and quiet), node y_0 has two children y_1 and y_2. Therefore, $n = 2$.

$$V_0 = \{\text{T2, T3, T4, T6, T7}\} \text{ at node } y_0$$

$$Y(1,0) = 3 \text{ (number of patterns in } V_0 \text{ of class } C_1)$$

$$Y(2,0) = 2 \text{ (number of patterns in } V_0 \text{ of class } C_2)$$

$$Z(0) = 5 \text{ (number of patterns in } V_0)$$

$$V_1 = \{\text{T2, T3, T7}\} \text{ at node } y_1, \text{ where HABIT} = \text{gabby}$$

$$Y(1,1) = 2 \text{ (number of patterns in } V_1 \text{ of class } C_1)$$

$$Y(2,1) = 1 \text{ (number of patterns in } V_1 \text{ of class } C_2)$$

$$Z(1) = 3 \text{ (number of patterns in } V_1)$$

$$V_2 = \{\text{T4, T6}\} \text{ at node } y_2, \text{ where HABIT} = \text{quiet}$$

$$Y(1,2) = 1 \text{ (number of patterns in } V_2 \text{ of class } C_1)$$

$$Y(2,2) = 1 \text{ (number of patterns in } V_2 \text{ of class } C_2)$$

$$Z(2) = 2 \text{ (number of patterns in } V_2)$$

$$I(V_0) = -\sum_{k=1}^{m} \frac{Y(k,0)}{Z(0)} \log \frac{Y(k,0)}{Z(0)} = 0.9710 \text{ [from Eqn (2.1)]}$$

$$I_{\text{HABIT}}(V_0) = -\sum_{j=1}^{n} \frac{Z(j)}{Z(0)} \sum_{k=1}^{m} \frac{Y(k,j)}{Z(j)} \log \frac{Y(k,j)}{Z(j)} = 0.9507 \text{ [from Eqn (2.2)]}$$

$$S_{\text{HABIT}}(V_0) = -\sum_{j=1}^{n} \frac{Z(j)}{Z(0)} \log \frac{Z(j)}{Z(0)} = 0.9710 \text{ [from Eqn (2.4)]}$$

$$G_{\text{HABIT}}(V_0) = \frac{I(V_0) - I_{\text{HABIT}}(V_0)}{S_{\text{HABIT}}(V_0)} = 0.0208 \text{ [from Eqn (2.5)]}$$

FIGURE 2.17 Evaluating $G_{\text{HABIT}}(V_0)$, the ratio of information gain of HABIT, at the right child of the root of a decision tree being built from the professor–student training set, FOOTWEAR having already been examined at the root, as in Fig. 2.10.

A default prule (see Section 1.1) is not needed because the decision tree shown in Fig. 2.9, from which the above prules are obtained, does not reject any patterns. The recall pattern R4 given in Fig. 1.2 (HABIT = quiet, EATS = roasted, and FOOTWEAR = clogs) is classified as student, by firing the first prule. This classification is the same as that given by the decision tree shown in Fig. 2.9.

In general, a decision tree and the prules obtained from that tree should provide the same results for a given pattern. Using prules is usually preferred because the prules present the knowledge used in classification in a form that is easy to read and understand. For the learning program, building a decision tree from a given training set thus becomes a step in obtaining prules for that set.

You may want to compare the above prules obtained from Fig. 2.9 with those obtained by the SpecToGen procedure (Fig. 1.3) and those obtained by the GenToSpec procedure (Fig. 1.5).

Summary

A decision tree built from a training set can later be used to classify training and recall patterns. Since we can build multiple decision trees from a given training set, we select a decision tree, in which, at every non-leaf node, the attribute examined maximizes the ratio of information gain. The prules obtained from the decision tree built can be used for classification. Overall, a learning program should build the appropriate decision tree and then, if needed, obtain prules from it.

Exercises

1. In a programming language of your choice, implement the procedure to build decision trees described in this chapter. The user should be able to specify the criterion that will be used to select the attribute to be examined at a non-leaf node: (a) arbitrarily, (b) by maximizing $g_A(V_0)$, the gain in information defined in Eqn (2.3), or (c) by maximizing $G_A(V_0)$, the ratio of information gain defined in Eqn (2.5).

2. In the procedure to build decision trees (Section 2.2), the set V_i associated with a node x_i is split into subsets V_{i1}, V_{i2}, ... , V_{in}, which are then associated with nodes x_{i1}, x_{i2}, ... , x_{in}. Then each

of the subsets V_{i1}, V_{i2}, ..., V_{in} are similarly split. Can this be done recursively? If so, in a programming language of your choice, implement a procedure that uses recursion to build decision trees. The user should be able to specify the criterion for selecting the attribute to be examined at a non-leaf node as in the first exercise.

3. Write a program in a language of your choice that, given a decision tree built by the first or second exercise above, will be able to classify a given pattern (see Section 2.3).

4. Obtain prules from each of the decision trees shown in Figs 2.3 to 2.8 and Fig. 2.10; the prules from Fig. 2.9 have already been given in Section 2.6. Classify patterns of the recall set shown in Fig. 1.2 using each of these seven knowledge bases (the sets of prules). Display your results in a table. Check whether your results match those given in Fig. 2.11, which were obtained by classifying the recall patterns directly from the corresponding decision trees. Is any set of prules obtained from the decision trees identical to either of the sets of prules given in Figs 1.3 or 1.5, which were obtained, respectively, by the SpecToGen and GenToSpec procedures of Sections 1.4 and 1.5?

5. Discuss the advantages and disadvantages of classifying patterns directly from decision trees versus first obtaining prules from the trees and then classifying the patterns using the prules. Remember, in domains of practical interest, decision trees may have depths of 20 or more; and a knowledge base may have many hundred prules, if not more.

6. The example solved in Section 2.5 maximizes $G_A(V_0)$, the ratio of information gain shown in Eqn (2.5), to build the decision tree from the professor–student training set given in Fig. 1.1. Would you build the same decision tree if you, instead, maximized $g_A(V_0)$, the gain in information shown in Eqn (2.3)? If not, which tree would you build? In maximizing $g_A(V_0)$, do you need to evaluate $I(V_0)$, or can that be avoided? If so, why and how can it be avoided?

7. Let the set of attributes consist of HABIT, EATS, FOOTWEAR, and NAME for the professor–student training set given in Fig. 1.1. Build a decision tree by each of these two methods.

   (a) Maximize $G_A(V_0)$, the ratio of information gain
   (b) Maximize $g_A(V_0)$, the gain in information

Compare the results obtained by the two methods. Which method would you recommend? Why?

8. To maximize the ratio of information gain (Section 2.5), all equations require log to the base 2. Many calculators and programming languages do not have functions built in to evaluate logs to the base 2. However, they do have logs to the base 10 and to the base e. We can find logs to the base 2 by either of the following equations:

$$\log_2 y = 3.3219 \times \log_{10} y$$
$$\log_2 y = 1.4427 \times \log_e y$$

If we were to use either logs to the base 10 or logs to the base e to maximize the ratio of information gain in selecting the attribute to be examined, will we get different results? Discuss.

9. As you may be aware, a deck of playing cards has four suits: clubs, spades, diamonds, and hearts. The colour of clubs and spades is black, and that of diamonds and hearts is red. Each suit has 13 cards: cards one to ten are called *numbered* cards, and cards jack, queen, and king are called *court* or *face* cards. Build a decision tree from the training set given in Fig. 2.18, and then identify the classes of the following cards or reject them: king of hearts, three of diamonds, jack of clubs, queen of diamonds, six of spades, nine of hearts, king of spades, and five of clubs.

10. Measure twenty computer science departments that carry out both teaching and research, on the following attributes (the students are

NAME of card	Attributes		Class
	COLOUR	TYPE	
Jack of hearts	Red	Court	C_1
Nine of spades	Black	Numbered	C_2
King of clubs	Black	Court	C_2
Queen of diamonds	Red	Court	C_1
Six of clubs	Black	Numbered	C_2

FIGURE 2.18 Training set for playing cards.

postgraduate—Master's and doctoral—students, the faculty are those who teach these students and supervise their theses, and averages are over the last five years for the department).

(a) Average number of students in the department

(b) Average number of faculty in the department

(c) Average age of the faculty in the department (this is considered because the older faculty may have more experience)

(d) Average number of research papers published in journals every year by the department faculty and students

(e) Average monetary value (rupees, dollars, francs, or whatever be the currency unit in your country) of research and industrial grants received by the department every year

(f) Average number of doctoral theses accepted every year

Allocate ten departments into class C_1 and the remaining ten into class C_2, the C_1 departments being intuitively considered to be better than the C_2 departments. Arbitrarily, select seven departments from C_1 and seven from C_2 to form a training set. Put the remaining three departments each from C_1 and C_2 into a recall set. From a decision tree, develop prules to classify patterns of the training set into classes C_1 and C_2. Test the prules on the patterns of the recall set and report how many of the recall patterns were classified correctly. You may collect data for this exercise from what you know about computer science departments in some universities, or you may thoughtfully create data on your own. Solve this exercise doing all the calculations manually, or use the programs you wrote for the first three exercises.

Learning Objectives

This chapter contains the following topics:

- a training set containing patterns with missing attribute values
- developing the mathematical model for the ratio of information gain provided by an attribute that has missing values
- building a decision tree when some of the training patterns have missing attribute values
- associating each non-leaf node with an attribute being examined, and each leaf node with the probability of a pattern belonging to a particular class or being rejected
- classifying a given pattern using a decision tree built from training patterns with missing attribute values
- predicting the error rate on recall sets based on the error rate observed on the training set
- pruning a decision tree based on the predicted error rate
- windowing of training sets in an attempt to reduce the computation required in building a decision tree

3

Decision Trees: Extensions

3.1 Missing Attribute Values

At times, a few attribute values are missing from some of the patterns in a training set. Say, in an effort to improve the treatments for different kinds of cancer, training set data are collected from many hospitals about their cancer patients: the attributes for each patient may describe personal information about the patient, the symptoms the patient had, the nature of cancer, the treatment received, and the effects of the treatment. Suppose, owing to oversights in recording the data at the hospital, some attribute values are missing. For example, for one patient, the age—which may be considered to decide the kind of treatment to be prescribed—might not have been recorded; for another patient, the size of the cancerous tumour might not have been recorded.

We can delete from the training set any patterns with missing attribute values, but if there are many such patterns, the training set could become small, reducing the generality of the decision tree built. Often the patterns with missing attribute values are allowed to remain in the training set. The equations to evaluate the attribute's ratio of information gain, described in Section 2.5, are then modified as follows.

Let C_1, C_2, ..., C_m be the $m \geq 1$ classes to which the training patterns belong. Suppose V_0, a subset of the training set V, is associated with a node y_0 of a decision tree we are building. We want to evaluate the ratio of information gain for attribute A at node y_0, but the value of attribute A is missing for some of the patterns in V_0.

If attribute A has $n > 1$ possible values, it will split V_0 into sets V_1, V_2, ..., V_n. Those patterns in V_0 whose value of attribute A is missing will not be in any of the sets V_1 to V_n. We will temporarily put such patterns in a set V_{n+1}. For $1 \leq k \leq m$ and $0 \leq i \leq n$, let $Y(k,i)$ be the number of training patterns of class C_k in the set V_i. For $0 \leq i \leq (n+1)$, let $Z(i)$ be the number of patterns in V_i. Moreover, let

$$Z'(0) = Z(0) - Z(n+1)$$

Out of $Z(0)$ patterns in set $V(0)$, we thus have $Z'(0)$ patterns for which the value of attribute A is known. Proceeding as we did for Eqn (2.1) in

Section 2.5, we obtain the entropy of the set V_0 as (remember all logs are to the base 2)

$$I(V_0) = -\sum_{k=1}^{m} \frac{Y(k,0)}{Z'(0)} \log \frac{Y(k,0)}{Z'(0)} \tag{3.1}$$

Similarly, as we did for Eqn (2.2) in Section 2.5, we obtain the entropy of attribute A for set V_0 as

$$I_A(V_0) = -\sum_{j=1}^{n} \frac{Z(j)}{Z'(0)} \sum_{k=1}^{m} \frac{Y(k,j)}{Z(j)} \log \frac{Y(k,j)}{Z(j)} \tag{3.2}$$

The patterns that are put in V_{n+1} (that is, those whose value of attribute A is missing) are not used in Eqns (3.1) and (3.2), because A does not provide any information about their class. Nonetheless, to obtain the gain in information caused by attribute A, we need to multiply the difference between $I(V_0)$ and $I_A(V_0)$ by $Z'(0)/Z(0)$, the probability of the value of attribute A being known in the set V_0. Accordingly, Eqn (2.3) of Section 2.5 is altered to the following:

$$g_A(V_0) = \frac{Z'(0)}{Z(0)} [I(V_0) - I_A(V_0)] \tag{3.3}$$

Attribute A has effectively split V_0 into $n+1$ sets, V_1 to V_{n+1}. Then, by adapting Eqn (2.4) of Section 2.5, the split information of attribute A becomes

$$S_A(V_0) = -\sum_{j=1}^{n+1} \frac{Z(j)}{Z(0)} \log \frac{Z(j)}{Z(0)} \tag{3.4}$$

When $Z'(0) = Z(0)$ (that is, there are no patterns with missing values of attribute A), Eqns (3.1) to (3.4) become identical to Eqns (2.1) to (2.4). In other words, Eqns (2.1) to (2.4) are special cases of Eqns (3.1) to (3.4), respectively. The *ratio of information gain* of attribute A for set V_0 is then, as in Eqn (2.5) of Section 2.5,

$$G_A(V_0) = \frac{g_A(V_0)}{S_A(V_0)} \tag{3.5}$$

We want to maximize this ratio of information gain. For a given non-leaf node in the decision tree, we calculate this ratio for each attribute in the

NAME of training pattern	Attributes			Class
	HABIT	EATS	FOOTWEAR	
T1	Gabby	Baked	Clogs	Student
T2	Gabby	Roasted	Sandals	Professor
T3	Gabby	Baked	Sandals	Student
T4	Quiet	Fried	Sandals	Professor
T5	Gabby	Fried	—	Student
T6	Quiet	Baked	Sandals	Student
T7	Gabby	Fried	Sandals	Professor
T8	Quiet	Fried	Clogs	Student

FIGURE 3.1 Training set for recognizing professors and students with the value of FOOTWEAR missing for pattern T5. This training set is adapted from the training set given in Fig. 1.1.

node's candidate set of attributes. While evaluating the ratio of information gain for each attribute in the candidate set, it may happen that some attributes have missing values. The attribute that has the highest ratio is then selected for examination at the node. This criterion of selection can be used in step 6 of the procedure to build decision trees, described in Section 2.2.

As an example, let us build the decision tree for the professor–student training set given in Fig. 3.1. On comparing the training sets given in Figs 1.1 and 3.1, one will notice that the two training sets are the same except that the former has no missing attribute values, whereas the latter has the value of attribute FOOTWEAR missing for pattern T5.

The candidate set of attributes that can be examined at the root consists of HABIT, EATS, and FOOTWEAR. The ratio of information gain for HABIT and EATS can be obtained from Figs 2.13 and 2.14, because their values are the same in the training sets given in Figs 1.1 and 3.1. Accordingly, $G_{\text{HABIT}}(V_0) = 0.0047$ and $G_{\text{EATS}}(V_0) = 0.3233$. Since there exists a pattern with a missing value of FOOTWEAR, we need to use Eqns (3.1) to (3.5), to evaluate $G_{\text{FOOTWEAR}}(V_0)$. Figure 3.2 shows how FOOTWEAR splits those training patterns whose values of FOOTWEAR are known. The ratio

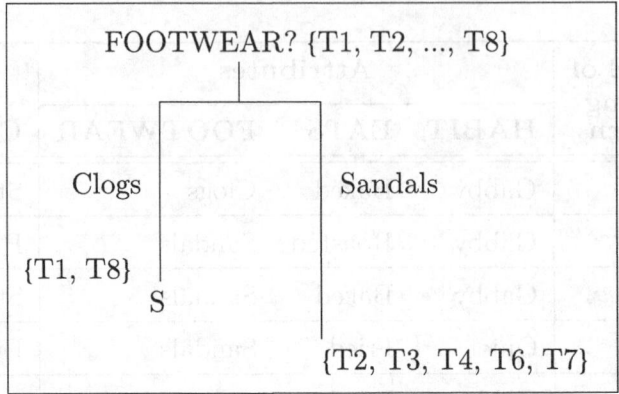

FOOTWEAR? {T1, T2, ..., T8}

Clogs   Sandals

{T1, T8}
          S

                    {T2, T3, T4, T6, T7}

FIGURE 3.2   Examining FOOTWEAR at the root of a decision tree to be built for the professor–student training set given in Fig. 3.1, in which the value of FOOTWEAR is missing for pattern T5. Hence, T5 is not associated with either of the children of the root. Both the patterns associated with the left child of the root belong to the class student, and hence the node becomes a leaf node with label S. Of the patterns associated with the right child of the root, patterns T2, T4, and T7 belong to the class professor; and patterns T3 and T6 belong to the class student.

of information gain for FOOTWEAR, $G_{\text{FOOTWEAR}}(V_0) = 0.1963$, has been calculated in Fig. 3.3. Among $G_{\text{HABIT}}(V_0)$, $G_{\text{EATS}}(V_0)$ and $G_{\text{FOOTWEAR}}(V_0)$, we notice that $G_{\text{EATS}}(V_0)$ is the largest. Hence, we select the attribute EATS to be examined at the root. At the root, the decision tree will, therefore, look like the tree in Fig. 3.4.

From the explanation in the caption of Fig. 3.4, we see that the left and the right children of the root are leaf nodes. The middle child of the root is a non-leaf node, and its candidate set of attributes consists of HABIT and FOOTWEAR. From Figs 3.5 and 3.6 we see that, at the middle child of the root, $G_{\text{FOOTWEAR}}(V_0) = 0.6351$ and $G_{\text{HABIT}}(V_0) = 0$. Since $G_{\text{FOOTWEAR}}(V_0)$ is the larger of the two ratios of information gain, we select FOOTWEAR to be examined at the middle child of the root of Fig. 3.4. The nodes generated by expanding the root's middle child are all leaf nodes, and we obtain a decision tree with a structure similar to the tree in Fig. 2.7, except that pattern T5 will not be associated with the left child of the node at which FOOTWEAR is examined, since the value of FOOTWEAR for pattern T5 is missing. Nonetheless, we will soon see a fraction of pattern T5 being associated with the left child.

To evaluate $G_{\text{FOOTWEAR}}(V_0)$ at the root of a decision tree for the professor–student training set given in Fig. 3.1, in which the value of FOOTWEAR is missing for pattern T5, the following procedure is used. The root is named y_0. The classes are C_1 (professor) and C_2 (student). Therefore, $m = 2$. Since FOOTWEAR has two possible values (clogs and sandals), root y_0 has two children y_1 and y_2. Therefore, $n = 2$. To understand the following material easily, you may want to look at Fig. 3.2, since it examines FOOTWEAR at the root.

$V_0 = \{\text{T1, T2, \ldots, T8}\}$ at node y_0

$Y(1,0) = 3$ (no. of known FOOTWEAR-value patterns in V_0 of class C_1)

$Y(2,0) = 4$ (no. of known FOOTWEAR-value patterns in V_0 of class C_2)

$Z(0) = 8$ (no. of patterns in V_0)

$Z'(0) = 7$ (no. of known FOOTWEAR-value patterns in V_0)

$V_1 = \{\text{T1, T8}\}$ at node y_1, where FOOTWEAR = clogs

$Y(1,1) = 0$ (no. of patterns in V_1 of class C_1)

$Y(2,1) = 2$ (no. of patterns in V_1 of class C_2)

$Z(1) = 2$ (no. of patterns in V_1)

$V_2 = \{\text{T2, T3, T4, T6, T7}\}$ at node y_2, where FOOTWEAR = sandals

$Y(1,2) = 3$ (no. of patterns in V_2 of class C_1)

$Y(2,2) = 2$ (no. of patterns in V_2 of class C_2)

$Z(2) = 5$ (no. of patterns in V_2)

$V_3 = \{\text{T5}\}$ (patterns with missing value of FOOTWEAR)

$Z(3) = 1$ (no. of patterns with missing value of FOOTWEAR)

$$I(V_0) = -\sum_{k=1}^{m} \frac{Y(k,0)}{Z'(0)} \log \frac{Y(k,0)}{Z'(0)} = 0.9852 \quad \text{[from Eqn (3.1)]}$$

$$I_{\text{FOOTWEAR}}(V_0) = -\sum_{j=1}^{n} \frac{Z(j)}{Z'(0)} \sum_{k=1}^{m} \frac{Y(k,j)}{Z(j)} \log \frac{Y(k,j)}{Z(j)} = 0.6936 \quad \text{[from Eqn (3.2)]}$$

$$g_{\text{FOOTWEAR}}(V_0) = \frac{Z'(0)}{Z(0)} [I(V_0) - I_{\text{FOOTWEAR}}(V_0)] = 0.25515 \quad \text{[from Eqn (3.3)]}$$

$$S_{\text{FOOTWEAR}}(V_0) = -\sum_{j=1}^{n+1} \frac{Z(j)}{Z(0)} \log \frac{Z(j)}{Z(0)} = 1.3 \quad \text{[from Eqn (3.4)]}$$

$$G_{\text{FOOTWEAR}}(V_0) = \frac{g_{\text{FOOTWEAR}}(V_0)}{S_{\text{FOOTWEAR}}(V_0)} = 0.1963 \quad \text{[from Eqn (3.5)]}$$

FIGURE 3.3 Evaluating $G_{\text{FOOTWEAR}}(V_0)$, the ratio of information gain for FOOTWEAR, at the root of a decision tree being built from the professor–student training set given in Fig. 3.1, in which the value of FOOTWEAR is missing for pattern T5, and hence Eqns (3.1) to (3.5) are used above. Set V_0 comprises the eight training patterns. The value of $G_{\text{FOOTWEAR}}(V_0)$ above is less than the value obtained in Fig. 2.15, where no value of FOOTWEAR was missing.

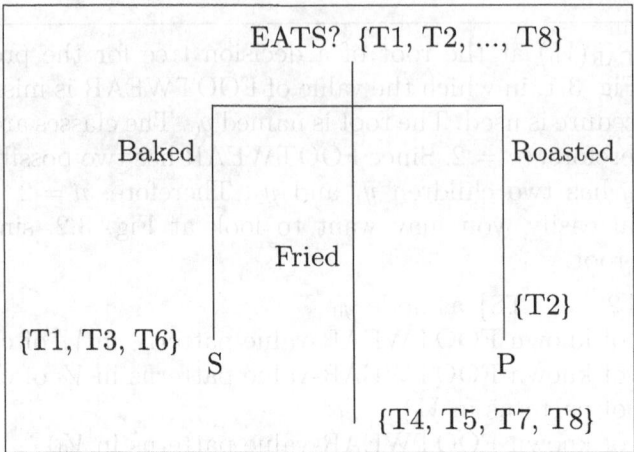

FIGURE 3.4 By maximizing the ratio of information gain, EATS has been selected for examination at the root of a decision tree being built for the professor–student training set in which the value of FOOTWEAR is missing for pattern T5. The three patterns associated with the left child of the root belong to the class student, and hence the node becomes a leaf node with label S. Of the patterns associated with the middle child of the root, patterns T4 and T7 belong to the class professor, and patterns T5 and T8 belong to the class student. To build this tree further, we would need to examine HABIT or FOOTWEAR at the middle child of the root. The single pattern associated with the right child of the root belongs to the class professor, and hence the node becomes a leaf node with label P.

Missing attribute values can cause a fraction of a pattern to appear in a set. Let the weight of a pattern in a set be equal to fraction of the pattern in the set. If the full pattern appears in the set, then its weight is 1. The weight of a pattern in a set can be said to reflect the probability that the pattern belongs to that set.

If an attribute A is being examined at a node y_0, with which set V_0 is associated, and if we are to do as described above, then patterns whose value of attribute A is missing in V_0 will not appear in any of the sets $V_1, V_2, \ldots, V_n$, into which A splits the set V_0. That will result in patterns with the missing values of attribute A not being associated with any of the descendants of y_0. So, in practice, a fraction of each such pattern is added to the set associated with each child of y_0, as described below.

1. For $1 \leq i \leq n$, let w_i be equal to the sum of the weights of the training patterns in set V_i.

To evaluate $G_{\text{FOOTWEAR}}(V_0)$ at the middle child of the root of the decision tree given in Fig. 3.4, in which EATS has been examined at the root, the following procedure is used. The value of FOOTWEAR is missing for pattern T5. The middle child of the root is named y_0. The classes are C_1 (professor) and C_2 (student). Therefore, $m = 2$. Since FOOTWEAR has two possible values (clogs and sandals), node y_0 has two children y_1 and y_2. Therefore, $n = 2$.

$V_0 = \{\text{T4, T5, T7, T8}\}$ at node y_0
$Y(1,0) = 2$ (no. of known FOOTWEAR-value patterns in V_0 of class C_1)
$Y(2,0) = 1$ (no. of known FOOTWEAR-value patterns in V_0 of class C_2)
$Z(0) = 4$ (no. of patterns in V_0)
$Z'(0) = 3$ (no. of known FOOTWEAR-value patterns in V_0)
$V_1 = \{\text{T8}\}$ at node y_1, where FOOTWEAR = clogs
$Y(1,1) = 0$ (no. of patterns in V_1 of class C_1)
$Y(2,1) = 1$ (no. of patterns in V_1 of class C_2)
$Z(1) = 1$ (no. of patterns in V_1)
$V_2 = \{\text{T4, T7}\}$ at node y_2, where FOOTWEAR = sandals
$Y(1,2) = 2$ (no. of patterns in V_2 of class C_1)
$Y(2,2) = 0$ (no. of patterns in V_2 of class C_2)
$Z(2) = 2$ (no. of patterns in V_2)
$V_3 = \{\text{T5}\}$ (patterns with missing value of FOOTWEAR)
$Z(3) = 1$ (no. of patterns with missing value of FOOTWEAR)

$$I(V_0) = -\sum_{k=1}^{m} \frac{Y(k,0)}{Z'(0)} \log \frac{Y(k,0)}{Z'(0)} = 0.9183 \quad \text{[from Eqn (3.1)]}$$

$$I_{\text{FOOTWEAR}}(V_0) = -\sum_{j=1}^{n} \frac{Z(j)}{Z'(0)} \sum_{k=1}^{m} \frac{Y(k,j)}{Z(j)} \log \frac{Y(k,j)}{Z(j)} = 0 \quad \text{[from Eqn (3.2)]}$$

$$g_{\text{FOOTWEAR}}(V_0) = \frac{Z'(0)}{Z(0)}[I(V_0) - I_{\text{FOOTWEAR}}(V_0)] = 0.9183 \quad \text{[from Eqn (3.3)]}$$

$$S_{\text{FOOTWEAR}}(V_0) = -\sum_{j=1}^{n+1} \frac{Z(j)}{Z(0)} \log \frac{Z(j)}{Z(0)} = 1.4466 \quad \text{[from Eqn (3.4)]}$$

$$G_{\text{FOOTWEAR}}(V_0) = \frac{g_{\text{FOOTWEAR}}(V_0)}{S_{\text{FOOTWEAR}}(V_0)} = 0.6351 \quad \text{[from Eqn (3.5)]}$$

FIGURE 3.5 Evaluating $G_{\text{FOOTWEAR}}(V_0)$, the ratio of information gain for FOOTWEAR, at the middle child of the root of the decision tree given in Fig. 3.4, in which EATS has been examined at the root. The value of FOOTWEAR is missing for pattern T5, and hence Eqns (3.1) to (3.5) are used above.

To evaluate $G_{\text{HABIT}}(V_0)$ at the middle child of the root of the decision tree given in Fig. 3.4, in which EATS has been examined at the root, the following procedure is used. The middle child of the root is named y_0. The classes are C_1 (professor) and C_2 (student). Therefore, $m = 2$. Since HABIT has two possible values (gabby and quiet), node y_0 has two children y_1 and y_2. Therefore, $n = 2$.

$$V_0 = \{\text{T4, T5, T7, T8}\} \text{ at node } y_0$$

$$Y(1,0) = 2 \text{ (no. of patterns in } V_0 \text{ of class } C_1)$$

$$Y(2,0) = 2 \text{ (no. of patterns in } V_0 \text{ of class } C_2)$$

$$Z(0) = 4 \text{ (no. of patterns in } V_0)$$

$$V_1 = \{\text{T5, T7}\} \text{ at node } y_1, \text{ where HABIT } = \text{gabby}$$

$$Y(1,1) = 1 \text{ (no. of patterns in } V_1 \text{ of class } C_1)$$

$$Y(2,1) = 1 \text{ (no. of patterns in } V_1 \text{ of class } C_2)$$

$$Z(1) = 2 \text{ (no. of patterns in } V_1)$$

$$V_2 = \{\text{T4, T8}\} \text{ at node } y_2, \text{ where HABIT } = \text{quiet}$$

$$Y(1,2) = 1 \text{ (no. of patterns in } V_2 \text{ of class } C_1)$$

$$Y(2,2) = 1 \text{ (no. of patterns in } V_2 \text{ of class } C_2)$$

$$Z(2) = 2 \text{ (number of patterns in } V_2)$$

$$I(V_0) = -\sum_{k=1}^{m} \frac{Y(k,0)}{Z(0)} \log \frac{Y(k,0)}{Z(0)} = 1.0 \text{ [from Eqn (2.1)]}$$

$$I_{\text{HABIT}}(V_0) = -\sum_{j=1}^{n} \frac{Z(j)}{Z(0)} \sum_{k=1}^{m} \frac{Y(k,j)}{Z(j)} \log \frac{Y(k,j)}{Z(j)} = 1.0 \text{ [from Eqn (2.2)]}$$

$$S_{\text{HABIT}}(V_0) = -\sum_{j=1}^{n} \frac{Z(j)}{Z(0)} \log \frac{Z(j)}{Z(0)} = 1.0 \text{ [from Eqn (2.4)]}$$

$$G_{\text{HABIT}}(V_0) = \frac{I(V_0) - I_{\text{HABIT}}(V_0)}{S_{\text{HABIT}}(V_0)} = 0 \text{ [from Eqn (2.5)]}$$

FIGURE 3.6　Evaluating $G_{\text{HABIT}}(V_0)$, the ratio of information gain of HABIT, at the middle child of the root of the decision tree given in Fig. 3.4, in which EATS has been examined at the root. No value of HABIT is missing, hence Eqns (2.1), (2.2), (2.4), and (2.5) are used above.

2. The sum of the weights is calculated as

$$\text{sum\_of\_weights} = \sum_{i=1}^{n} w_i$$

3. For every training pattern T in V_0 whose value of attribute A is missing, do steps 3.1 and 3.2.

3.1. Let f_0 be the fraction of pattern T present in V_0.

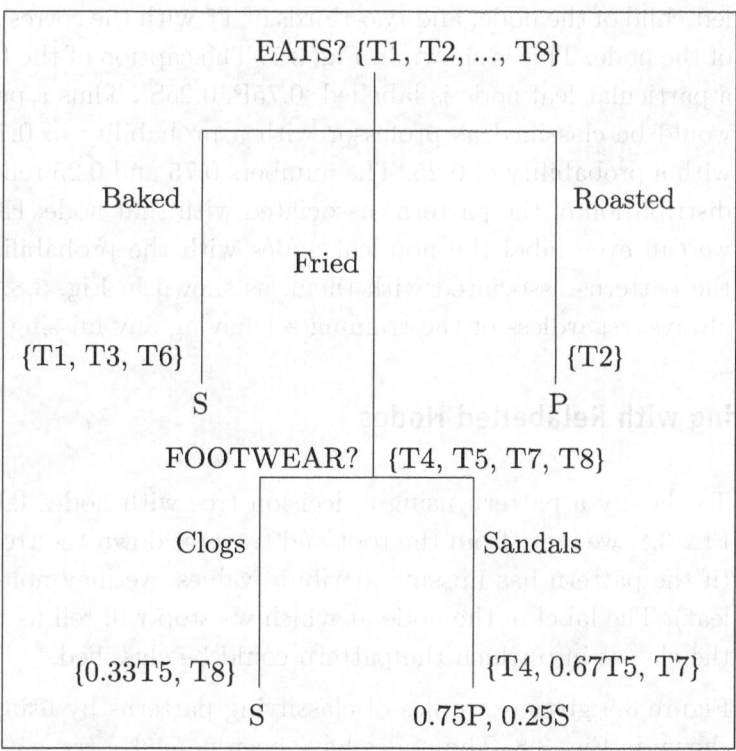

FIGURE 3.7   The decision tree showing the set of patterns associated with each node for the professor–student training set given in Fig. 3.1, in which the value of FOOTWEAR is missing for pattern T5. One-third (that is, 0.33) of T5 is associated with the left child of the node at which FOOTWEAR is examined. The left child is labelled S because the 1.33 patterns of T5 and T8 associated with it belong to the class student. Two-thirds (that is, 0.67) of T5 is associated with the right child of the node at which FOOTWEAR is examined. The right child is labelled '0.75P, 0.25S' because, of the 2.67 patterns associated with the child, two patterns (T4 and T7) belong to the class professor (2/2.67 = 0.75); and 0.67 pattern (namely T5) belongs to the class student (0.67/2.67 = 0.25). A pattern at this node will be classified as professor with a probability of 0.75, and as student with a probability of 0.25. You may want to compare this tree with the tree given in Fig. 2.7.

3.2. For $j = 1, 2, \ldots, n$ do step 3.2.1.

3.2.1. Add to set V_j the f_jth fraction of pattern T, where

$$f_j = \frac{w_j}{\text{sum\_of\_weights}} f_0$$

4. Return from the procedure.

If we are to apply this procedure to the node at which FOOTWEAR is examined in Fig. 3.4, then one-third of pattern T5 will be associated with the left child of the node, and two-thirds of T5 with the corresponding right child of the node. This is shown in Fig. 3.7. The caption of the figure explains how a particular leaf node is labelled '0.75P, 0.25S'. Thus a pattern at this node would be classified as professor with a probability of 0.75, and as student with a probability of 0.25. The numbers 0.75 and 0.25 reflect the probability distribution of the patterns associated with that node. Extending this idea, we can even label the non-leaf nodes with the probability distributions of the patterns associated with them, as shown in Fig. 3.8. This can be done always, regardless of the training set having any missing attribute values.

3.2   Classifying with Relabelled Nodes

To classify a pattern using a decision tree with nodes labelled as shown in Fig. 3.8, we start from the root and traverse down the tree as far as possible (if the pattern has missing attribute values, we may not be able to reach a leaf). The label of the node at which we stop will tell us the probabilities of the classes into which the pattern could be classified.

Figure 3.9 gives examples of classifying patterns by using the decision tree shown in Fig. 3.8. The nodes have been named $x, x_1, x_2, \ldots$ in the decision tree to make the figure easier to understand. The figure contains all the training patterns given in Fig. 3.1 and a few recall patterns. The figure shows that the classifications of few of the training patterns by the decision tree given in Fig. 3.8 are not the same as that given in Fig. 3.1.

Three patterns—R11, R12 and R13—given in Fig. 3.9 need comments, because for each of them we are unable to traverse down the tree: we have to stop at the root. In R11, the only attribute value known is FOOTWEAR equal to clogs, so from the label of the root x, we say that the pattern belongs to the class professor with a probability of 0.375, and it belongs to the class student with a probability of 0.625. In a situation, where we have to stop at the root itself, an alternative method can be used to classify the pattern. In the training set given in Fig. 3.1, two patterns, T1 and T8, have FOOTWEAR equal to clogs, and both belong to the class student. So we recognize pattern R11 as belonging to the class student.

Similarly, for pattern R12 (HABIT = gabby), we notice in the training set given in Fig. 3.1, that out of the five patterns with HABIT equal to gabby, two (T2 and T7) belong to the class professor, and three (T1, T3, T5) belong to the class student. Accordingly, we recognize pattern R12 as belonging to the class professor with probability 0.4, and student with probability 0.6. Using this method to classify patterns, such as R11 and R12, requires that

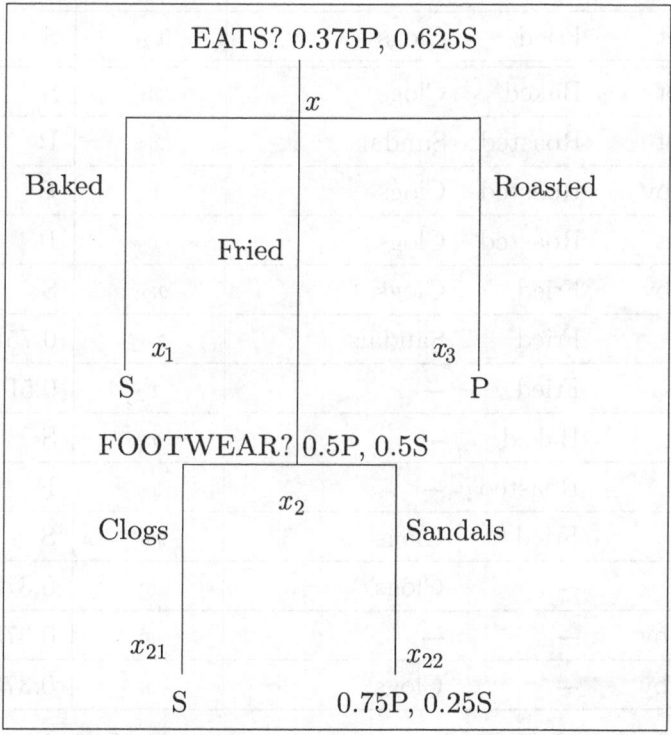

FIGURE 3.8   The decision tree obtained from Fig. 3.7 with each node labelled with the probability distribution of the classes of the patterns associated with that node. The tree is for the professor–student training set given in Fig. 3.1, in which the value of FOOTWEAR is missing for pattern T5. Once we have labelled the nodes as above, we need not show the patterns associated with each node. The nodes have been named $x, x_1, x_2, \ldots$ to make Fig. 3.9 easier to understand.

NAME of pattern	Attributes			Node of Fig. 3.8 reached	Classification
	HABIT	EATS	FOOTWEAR		
T1	Gabby	Baked	Clogs	x_1	S
T2	Gabby	Roasted	Sandals	x_3	P
T3	Gabby	Baked	Sandals	x_1	S
T4*	Quiet	Fried	Sandals	x_{22}	0.75P, 0.25S
T5*	Gabby	Fried	—	x_2	0.5P, 0.5S
T6	Quiet	Baked	Sandals	x_1	S
T7*	Gabby	Fried	Sandals	x_{22}	0.75P, 0.25S
T8	Quiet	Fried	Clogs	x_{21}	S
R1	Quiet	Baked	Clogs	x_1	S
R2	Quiet	Roasted	Sandals	x_3	P
R3	Gabby	Roasted	Clogs	x_3	P
R4	Quiet	Roasted	Clogs	x_3	P
R5	Gabby	Fried	Clogs	x_{21}	S
R6	—	Fried	Sandals	x_{22}	0.75P, 0.25S
R7	—	Fried	—	x_2	0.5P, 0.5S
R8	—	Baked	—	x_1	S
R9	—	Roasted	—	x_3	P
R10	—	Fried	Clogs	x_{21}	S
R11	—	—	Clogs	x	0.375P, 0.625S
R12	Gabby	—	—	x	0.375P, 0.625S
R13	Gabby	—	Clogs	x	0.375P, 0.625S

FIGURE 3.9　Classifications of some professor–student patterns using the decision tree given in Fig. 3.8. A missing value of an attribute is shown by a dash. Patterns T1 to T8 are the training patterns taken from Fig. 3.1. An asterisk besides the name of a training pattern indicates that its classification here is not identical to that in its training set given in Fig. 3.1. Patterns R1 to R13 are recall patterns, of which R1 to R4 are the same as those in Fig. 1.2. The classifications for patterns R11 to R13 have been obtained from the label at the root of the decision tree. For such patterns, an alternative method of classification is discussed in Section 3.2.

the class distributions for each attribute value and combinations of attribute values in the training set be available at the time of classification.

We would need the class distribution for attribute value combinations for pattern R13 (HABIT = gabby and FOOTWEAR = clogs). According to the training set given in Fig. 3.1, only one pattern, namely T1, has HABIT equal to gabby and FOOTWEAR equal to clogs; and it is classified as student, so we would recognize R13 also as belonging to the class student. The method (the one used in Fig. 3.9 or the one discussed here) you will use for patterns such as R11 to R13 is your design choice. As is clear, the two methods may not give the same results.

3.3 Error Rates on Recall Sets

Suppose class C is the most frequent in a set V' associated with a node in a decision tree. We can say that a pattern at that node will be classified into C, with *error rate*

$$e(V') = \frac{|V'| - \text{number of patterns of class } C \text{ in } V'}{|V'|}$$

where $|V'|$ is the number of patterns in V'. For instance, if $|V'| = 90$, and the number of patterns of class C in V' are 72, then the error rate is $(90 - 72)/90$, that is, 0.2. The error rate $e(V')$ gives us the estimated probability of error obtained in the training patterns in V'.

We can use statistics (only an overview of the statistics involved is described here to avoid distraction from the current discussion) to predict the range of error rate on recall sets provided all of the following conditions are true:

- The value of $e(V')$ is not too close to 0 or 1
- $[|V'| \times e(V')\{1 - e(V')\}] \geq 5$
- $|V'| \geq 30$ (the number of training patterns is usually large, hence this condition should hold often)
- The patterns in V' are independent of one another
- The recall sets will be drawn from the same population of patterns as V'

When the range of error rate for $0 \leq N \leq 100$ per cent of the recall sets is predicted, it is said to be prediction with N *per cent confidence.* Under

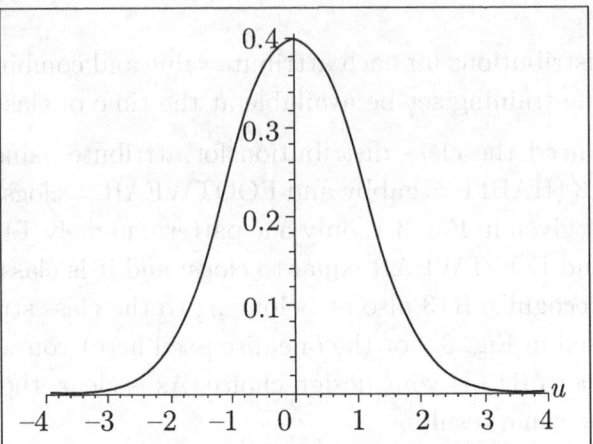

FIGURE 3.10 Normal distribution curve, obtained by plotting $\frac{1}{\sqrt{2\pi}}e^{-u^2/2}$.

the above conditions, the range of error rate can be approximated to have a Normal (or Gaussian) distribution (Fig. 3.10), which is plotted by the function

$$\frac{1}{\sqrt{2\pi}}e^{-u^2/2}$$

The area under the curve from $u = -\infty$ to $u = \infty$ is 1, denoting, for this purpose, 100 per cent confidence. For a given value of N per cent confidence, we want to find the value of z such that the area under the curve from $u = -z$ to $u = z$ is $N/100$. Since the curve is symmetric, this is equivalent to finding the value of z such that the area under the curve from $u = 0$ to $u = z$ is $N/200$. This can be done by solving for z in the equation

$$\frac{1}{\sqrt{2\pi}}\int_0^z e^{-u^2/2}du = \frac{N}{200}$$

We can write a computer program that, given a value of N, will find the value of z from the above equation. Alternatively, we can find the value of z corresponding to N from Fig. 3.11. Once we have selected a value of N and found the corresponding value of z, we can predict—according to a statistical result—that for N per cent of the recall sets, the error rate on a

$N(\%)$	z	$N(\%)$	z	$N(\%)$	z	$N(\%)$	z	$N(\%)$	z	$N(\%)$	z
0	0.00	17	0.21	34	0.44	51	0.69	68	1.00	85	1.44
1	0.01	18	0.23	35	0.45	52	0.71	69	1.02	86	1.48
2	0.03	19	0.24	36	0.47	53	0.73	70	1.04	87	1.51
3	0.04	20	0.25	37	0.49	54	0.74	71	1.06	88	1.56
4	0.05	21	0.27	38	0.50	55	0.76	72	1.08	89	1.60
5	0.06	22	0.28	39	0.51	56	0.77	73	1.10	90	1.64
6	0.08	23	0.29	40	0.52	57	0.79	74	1.13	91	1.70
7	0.09	24	0.31	41	0.53	58	0.81	75	1.15	92	1.75
8	0.10	25	0.32	42	0.55	59	0.82	76	1.18	93	1.81
9	0.11	26	0.33	43	0.57	60	0.84	77	1.20	94	1.88
10	0.13	27	0.34	44	0.58	61	0.86	78	1.23	95	1.96
11	0.14	28	0.36	45	0.60	62	0.88	79	1.25	96	2.05
12	0.15	29	0.37	46	0.61	63	0.90	80	1.28	97	2.17
13	0.17	30	0.39	47	0.63	64	0.92	81	1.31	98	2.33
14	0.18	31	0.40	48	0.64	65	0.93	82	1.34	99	2.58
15	0.19	32	0.41	49	0.66	66	0.96	83	1.37	99.99	3.90
16	0.20	33	0.43	50	0.67	67	0.97	84	1.41	100	∞

FIGURE 3.11 Values of z corresponding to confidence N varying from 0 to 100 per cent. For a given N per cent, the area under the normal distribution curve (Fig. 3.10) between $u = 0$ and $u = z$ is $N/200$. Thus, for a given N, the value of z can be obtained from the equation

$$\frac{1}{\sqrt{2\pi}} \int_0^z e^{-u^2/2} du = \frac{N}{200}.$$

given recall set will lie in the interval

$$e(V') \pm z\sqrt{\frac{e(V')[1 - e(V')]}{|V'|}}$$

Suppose we want to predict the error rate with 95 per cent confidence for the above example, where $|V'|$ is 90, and $e(V')$ is 0.2. We see from Fig. 3.11

that, for $N = 95$, the value of z is 1.96. Substituting these values in the above equation, we can predict that, for 95 per cent of the recall sets, the error rate will lie in the interval

$$0.2 \pm 1.96\sqrt{0.2(1 - 0.2)/90}$$
$$= 0.2 \pm 0.08$$
$$= [0.12, 0.28]$$

The square brackets, as is conventional in mathematics, represent a closed interval. From the above, we can make the following predictions.

- For 95 per cent of the recall sets, the error rate will be at least 0.12 and at the most 0.28.
- For 2.5 per cent of the recall sets (that is, with 2.5 per cent confidence), the error rate will be less than 0.12. Thus, for 97.5 per cent of the recall sets (that is, with 97.5 per cent confidence), the error rate will be 0.12 or more.
- For 2.5 per cent of the recall sets (that is, with 2.5 per cent confidence), the error rate will be more than 0.28. Thus, for 97.5 per cent of the recall sets (with 97.5 per cent confidence), the error rate will be 0.28 or less.

The upper bound of the interval [0.12, 0.28] can be considered to be the upper bound of the error rate on recall sets. Thus, the upper bound of the error rate for recall sets can be considered to be 0.28.

3.4 Pruning Decision Trees

In a decision tree, consider a node x_i whose children x_{i1} to x_{in} are leaf nodes. If the upper bound of the predicted error rate at x_i is lower than the upper bounds of the predicted error rates at each of its children x_{i1} to x_{in}, then—as a design choice—we can delete the children x_{i1} to x_{in}. This is known as *pruning* the tree. Node x_i then becomes a leaf node. We can repeat this process, thus iteratively pruning the decision tree. A pruned tree, being shallower and thus examining fewer attributes, is expected to classify patterns faster than an unpruned tree. Moreover, it has often been found in practice that pruned decision trees perform better on recall sets than unpruned trees: deep, unpruned decision trees tend to memorize the training set. They are said to be *overfitted* to the training set, that is, they become too specific to the training set. Pruned trees generalize better and so perform better on recall sets.

A node in a decision tree, be it unpruned or pruned, can be labelled with the class into which a given pattern will be classified at that node, together with the error rate $e(V')$ and the upper bound of the predicted error rate (if such prediction was made because the conditions listed in Section 3.3 were true at that node). That way, when the tree is used to classify some given pattern, the user of the decision tree can know not only the pattern's classification, but also the error rate $e(V')$ and the upper bound of the predicted error rate. All three—the classification, the error rate $e(V')$, and the upper bound of the error rate—are taken from the label of the node at which the pattern is classified. A user may choose to have acceptable error rates, depending on the domain of patterns—the acceptable error rates would understandably be lower for a medical domain than for recognizing, say, professors and students. For a given pattern, if any of the error rates provided by the decision tree is higher than what the user considers acceptable, then the user may consider the pattern rejected, that is, the decision tree has failed to recognize the pattern. It is the user who then determines when a pattern can be considered as rejected by the decision tree.

3.5 Windowing of Training Sets

For building decision trees, which perform well on recall patterns, large training sets are required. Large training sets, however, slow down the process of building the decision tree. One approach, which sometimes works in practice, is to select a proper subset W of V, the large training set. W is called the *window*. The class distribution of the patterns in W could be the same as in V, or the classes in W may be equiprobable.

Build a decision tree from W. Classify the patterns of $(V - W)$ with the decision tree built. If the error exceeds an acceptable limit, add the misclassified patterns to W. Rebuild the decision tree on the enlarged W. Repeat the above process till the error rate falls below the acceptable limit. This, of course, is not guaranteed to happen, and in the worst case, the full V will be needed to build the decision tree. One may choose to try building the decision tree by such windowing of the training set.

Summary

At times, patterns in a training set may have some of the attribute values missing. To build a decision tree from such a training set, the equation to maximize the ratio of information gain is modified. Such a tree can have fractions of patterns associated with a node. Each node can be labelled with the class probability distribution of the patterns associated with the node. From the distributions, we can get the error rate on the training patterns, which can be used to predict the error rates on recall sets. For very large training sets, decision trees can be built by windowing the training sets.

Exercises

1. Modify the program you wrote either in the first exercise or second exercise of Chapter 2 to build a decision tree with some attribute values missing. With the observed error on the training set, the program should be able to predict the range of error rates on recall sets at a given confidence level. For a given confidence level N, the program should read the value of z from Fig. 3.11 stored in computer memory, or it could calculate the value of z from the equation

    $$\frac{1}{\sqrt{2\pi}} \int_0^z e^{-u^2/2} du = \frac{N}{200}$$

    To find the value of z from the above equation, the knowledge of procedures for numerical integration and for finding roots of equations is required.

2. With the program you wrote in the first exercise, build a decision tree when the value of HABIT is missing from pattern T3 of the training set given in Fig. 1.1. Use the tree to classify the patterns listed in Fig. 3.9. Predict the range of error rates on recall sets at confidence levels of ninety, ninety-five, and ninety-nine per cent. Although the conditions required to predict error rates (listed in Section 3.3) will not be true, you may assume they are true for this exercise.

3. Repeat the second exercise when the value of HABIT is missing from pattern T3, and the value of EATS is missing from pattern T7 of the training set given in Fig. 1.1.

4. In a constituency of 200,000 voters, suppose you interview 1000 arbitrarily selected people and find out that, in the upcoming election, 53 per cent plan to vote for the Revolutionary party and the remaining for the Rotationary party. Based on Section 3.3, can you predict with 99 per cent confidence that, in the election, the percentage of votes for the Revolutionary party will lie in the interval [49, 57], and for the Rotationary party in the interval [43, 51]? If you had, instead, interviewed 10,000 people, and found that the same percentages planned to vote for each of the two parties, then with 99 per cent confidence find the corresponding two voting intervals. Have the intervals narrowed or widened? What would be the intervals if you carried out the last step at 95 per cent confidence, instead of 99 per cent?

Learning Objectives

This chapter contains the following topics:

- evolving through iterative refinement
- the criterion for the fitness of an antecedent
- binary coding of attribute values to represent an antecedent by a string of binary bits
- the operation of reproduction, which duplicates a given binary string
- the operation of mutation, which changes the bits in a binary string
- the operation of crossover, which produces new binary strings from a given pair of strings
- creating a random initial population of binarily represented, antecedents of the production rule for a class
- iteratively carrying out operations of reproduction, mutation, and crossover on the population of antecedents to obtain new antecedents
- obtaining antecedents with acceptable fitness by retaining antecedents with high fitness and discarding the rest

4

Obtaining Prules
by Evolution

4.1 Iterative Refinement

An architect may draw many sketches of a building being designed, each sketch different from the other. If some sketch does not fit the requirements at all, it is discarded. If a portion of a sketch fits the requirements partly, changes are made to that portion so that it fits better. Sometimes, a portion of one sketch is incorporated into another sketch so that the combined sketch fits closer to the requirements. Such discarding, changing, and incorporating may be repeated until a sketch that fits the requirements the best is developed. By such iterative refinement, the design of the building can be said to have *evolved* from the initial sketches.

4.2 Fitness of an Antecedent

To develop prules from a given training set, we can follow an approach similar to the architect's. Starting with some possible antecedents for prules to recognize a given class, we iteratively refine the antecedents so that they evolve to fit more and more training patterns.

The antecedent of a prule with class C in the consequent is defined to *fit* a training pattern under these conditions: (i) if the pattern belongs to class C, then the antecedent covers the pattern, that is, the attribute values of the pattern make the antecedent true; (ii) if the pattern does not belong to class C, then the antecedent does not cover the pattern. Alternatively, the antecedent fits a training pattern if the antecedent covers the pattern when it should, and does not cover the pattern when it should not. The *fitness* of the antecedent is then defined to be the proportion of training patterns the antecedent fits. Suppose we want to find the fitness of the antecedent of the following prule obtained from the professor–student training set given in Fig. 1.1.

> If HABIT = gabby,
> and EATS = fried or roasted,
> then the pattern belongs to the class professor.

Because the prule has professor in the consequent, the antecedent should cover the professor training patterns, and it should not cover the non-professor training patterns. Hence, the antecedent should cover professor patterns T2, T4, and T7; but, in fact, it covers patterns T2, T5, and T7. Thus the antecedent fits two professor patterns T2 and T7, since they are common to both the lists. Moreover, the antecedent should not cover non-professor patterns T1, T3, T5, T6, and T8; but, in fact, it does not cover T1, T3, T4, T6, and T8. Accordingly, the antecedent fits four non-professor patterns T1, T3, T6, and T8. Summing up, we can say that the antecedent of the prule fits six patterns T1, T2, T3, T6, T7, and T8. Since the antecedent fits six out of the eight training patterns, the fitness of the antecedent is $6/8 = 0.75$.

4.3   Binary Representation of an Antecedent

For evolving antecedents with high fitness, it is customary to represent the antecedents as binary strings. To do that, we need to first choose the binary code for each possible attribute value. Figure 4.1 shows the codes chosen for the attributes of the professor–student training set given in Fig. 1.1. We also need to choose the sequence in which we will write the attribute values in an antecedent: The sequence for attribute values here is in the order HABIT, EATS, and FOOTWEAR. Consider the prule given in Section 4.2.

If HABIT = gabby,
and EATS = fried or roasted,
then the pattern belongs to the class professor.

The antecedent of the prule does not examine FOOTWEAR, which means that the value of FOOTWEAR is immaterial, that is, it can have any of its two possible values (clogs or sandals). So, the prule can alternatively be written as

If HABIT = gabby,
and EATS = fried or roasted,
and FOOTWEAR = clogs or sandals,
then the pattern belongs to the class professor.

According to the codes chosen in Fig. 4.1, the antecedent of the above prule is then represented as

01 011 11

Attribute	Value	Binary Code
HABIT	= Gabby	01
	= Quiet	10
	= Gabby or quiet	11
	Impossible	00
EATS	= Fried	001
	= Roasted	010
	= Baked	100
	= Fried or roasted	011
	= Fried or baked	101
	= Roasted or baked	110
	= Fried or roasted or baked	111
	Impossible	000
FOOTWEAR	= Clogs	01
	= Sandals	10
	= Clogs or sandals	11
	Impossible	00

FIGURE 4.1 Binary codes chosen by the evolutionary procedure to represent attribute values of the professor–student training set given in Fig. 1.1.

The space between the values of the different attributes are only for visual clarity. The blank spaces should not be considered as a part of the binary string.

4.4 Reproduction, Mutation, and Crossover

During the evolutionary procedure, three operations are carried out on the binary strings, the operations of *reproduction, mutation,* and *crossover*, each of which is explained below.

By the operation of *reproduction,* a string is duplicated; for example, the string 1010100 is reproduced as 1010100.

By the operation of *mutation,* a bit in a string is changed: if the bit is 1, it is changed to 0; if it is 0, it is changed to 1. Consider the following strings, usually referred to as a *population* of strings:

```
10101001
00010110
11010111
00011001
01101011
```

Suppose we decide that the probability of a bit mutating is 0.001. For each bit, we invoke a uniform random number generator, which returns a real number between 0 and 1. If the number returned is 0.001 or less, we mutate the bit (1 to 0, or 0 to 1, as the case may be); otherwise, we let the bit remain as it is. Since the above population has a total of 40 bits, the probability of a mutation taking place is $40 \times 0.001 = 0.04$. If the population were to contain 1000 bits, then the probability of a mutation taking place would be $1000 \times 0.001 = 1$.

A random number generator, which will be referred to often in this chapter, is assumed to be a *uniform* random number generator implying that all real numbers between 0 and 1 are generated with equal probability. Most programming languages have a built-in software function to simulate a random number generator. Before a random number generator is invoked for the first time in a program, the generator will usually accept an initial value called the *seed* value. The sequence of random numbers generated by repeatedly invoking the generator then depends on the seed value given. If the same seed value is given every time the program is run, then the same sequence of random numbers will be generated. An alternative approach is to give different seed values whenever the program is run. A recommended seed value is some combination of the date and time at which the program is run. Therefore, the seed value will keep changing every time the program is run, thus providing different sequences of random numbers. The date and time can be obtained using the date and time functions available on the computer. If something is referred to as *randomly* chosen, it means it was chosen using a random number generator.

The third operation carried out on binary strings is *crossover*, in which two strings are used to produce other strings. Suppose we have two binary strings represented as β and δ, each of *length* b, that is, each string has b bits. Let us choose an integer value c, such that $1 \leq c < b$. Counting from the left, if the first c bits of β are represented by β_1, the remaining bits of β by β_2, the first c bits of δ by δ_1, and the remaining bits of δ by δ_2, then by carrying out crossover on the strings

$$\beta_1 \ \beta_2$$
$$\delta_1 \ \delta_2$$

we obtain the strings

$$\beta_1 \ \delta_2$$
$$\delta_1 \ \beta_2$$

The variable c is said to denote the *crossover site,* and the strings β and δ are said to be each other's *mates.* From two mates each of length b, by varying c from 1 to $b - 1$, we can obtain by crossover as many as $2(b - 1)$ strings. For instance, from the strings

1011011
0100110

we obtain the following strings by carrying out crossover, the crossover site being at 3:

1010110
0101011

Since the strings are of length 7, by varying the crossover site from 1 to 6, we can obtain as many as $2 \times 6 = 12$ strings. Crossover can be viewed as producing children strings from two parent strings.

4.5 Evolutionary Procedure for Prules

By the *evolutionary procedure* described below, we can obtain prules having a given class C in the consequent. One by one we can adopt the procedure for each of the classes in the training set.

Adopt the definition of the fitness of an antecedant given in Section 4.2.

1. Alternatively, modifying the definition of the fitness of an antecedent given in Section 4.2, one could define it as some function—say, the

square—of the proportion of the training patterns the antecedent fits. It is being assumed in the steps below that the maximum possible fitness of an antecedent is 1. Moreover, an antecedent fits all training patterns if, and only if, the antecedent's fitness is 1.

2. Choose the binary codes for attribute values (as shown in Fig. 4.1).

3. Choose a value for population size $P \geq 1$, that is, the population will have at the most P strings, each string denoting an antecedent, as explained in step 7 below.

4. Choose a value for $G \geq 1$, for the maximum number of times the operations of reproduction, mutation, and crossover will be carried out in a selected order. Each time these operations are carried out, a new *generation* of strings is said to be obtained. G is thus the maximum number of generations desired.

5. Choose a value for $F \leq 1$, such that antecedents with a fitness of F or more are acceptable. If you choose $F < 1$, it will mean that it is acceptable to use an antecedent in a prule for recognizing class C, although the antecedent may not fit all the training patterns.

6. $g \leftarrow 0$
   Initialize variable g to zero so it can count the generations.

7. Using the binary codes selected in step 2, create an initial population of P antecedents, where P is the value chosen in step 3. Each antecedent is for a prule whose consequent will assign a pattern to class C. The antecedents may be created randomly, or they may be based on intuition and analysis of the training set.

8. If the population has an antecedent in which the code for any attribute value is impossible (for instance, code 00 for HABIT in Fig. 4.1), then delete the antecedent, since the antecedent will always be false. Alternatively, as a design choice, one or more of the bits in the impossible attribute value can be randomly changed—from 0 to 1, or 1 to 0—so that the attribute value is no longer impossible.

9. Calculate the fitness for each antecedent in the population, as described in Section 4.2.

10. If the population has more than P antecedents, P being the value chosen in step 3, then retain the P antecedents with the highest fitness, and delete the remaining.

11. Put each antecedent whose fitness is $\mathcal{F}$ or more in a separate prule whose consequent assigns a pattern to class C. This will develop one or more prules for recognizing class C.

12. If no more prules are required to recognize class C, then return with the prules developed till now.

13. If $g > \mathcal{G}$, that is, the number of generations has exceeded the limit chosen in step 4, return with all the prules developed till now. (If no prules are returned, then you may invoke the procedure again by changing one or more of these: population size $\mathcal{P}$ in step 3, maximum number of generations $\mathcal{G}$ in step 4, acceptable fitness $\mathcal{F}$ in step 5, or the initial random antecedents in step 7.)

14. $g \leftarrow g + 1$
    Increment the counter for generations by 1.

15. Carry out reproduction, mutation, and crossover in a chosen order to obtain the next generation of antecedents. In reproduction, the higher the fitness of an antecedent, the more likely it should be reproduced. In mutation, for every bit, a random number generator is invoked to decide whether the bit should be mutated. In crossover, a random number generator is invoked to decide which antecedents will be each other's mates, and what will be the crossover site.

16. Go to step 8.

As an example, Figs 4.2 to 4.5 show the evolutionary development of a prule to recognize professors of the professor–student training set given in Fig. 1.1. The caption of each figure explains the example step by step. In Fig. 4.5, we obtain the antecedent 11 011 10, with a fitness of 1, since it fits all training patterns. The antecedent can thus be used in a prule for recognizing professors. Translating the binary representation of this antecedent into words, as described in Section 4.3, we obtain the following prule.

> If EATS = fried or roasted,
> and FOOTWEAR = sandals,
> then the pattern belongs to the class professor.

Because of the 'or' in the first condition of the antecedent, the prule can be split into the two prules

Ante-cedent number i	Antecedent	Training patterns that the antecedent fits	Fitness F_i	Probability of reproduction $\dfrac{F_i}{\sum\limits_{j=1}^{4} F_j}$
1	11 001 01	T1, T3, T6	$3/8 = 0.375$	$0.375/2.125 = 0.176$
2	11 101 01	T3, T6	$2/8 = 0.250$	$0.250/2.125 = 0.118$
3	01 011 11	T1, T2, T3 T6, T7, T8	$6/8 = 0.750$	$0.750/2.125 = 0.353$
4	01 111 10	T1, T2, T5 T6, T7, T8	$6/8 = 0.750$	$0.750/2.125 = 0.353$
			$\sum\limits_{i=1}^{4} F_i = 2.125$	

FIGURE 4.2   An example of binarily represented initial antecedents to be used in prules for recognizing professors of the professor–student training set given in Fig. 1.1. Four antecedents were chosen in the population of antecedents. The four antecedents above were randomly obtained; that is, for each bit in an antecedent, if the random number generated was 0.5 or less, then a 0 bit was put in the antecedent; otherwise, a 1 bit. The operation of reproduction was carried out first. For that, each antecedent was first translated into words, as discussed in Section 4.3; for instance, antecedent number 1 in words is 'EATS = fried and FOOTWEAR = clogs'. The fitness of each antecedent was calculated as described in Section 4.2. The relative fitness of each antecedent was considered to be the probability of its reproduction. Thus, the higher the fitness of an antecedent, the more likely it was to be reproduced. To have four antecedents after reproduction, a random number generator was invoked four times. Each time: if the number returned was less than or equal to 0.176, then the first antecedent was reproduced; if the number returned was more than 0.176, but less than or equal to $0.176 + 0.118 = 0.294$, then the second antecedent was reproduced; if the number returned was more than 0.294, but less than or equal to $0.294 + 0.353 = 0.647$, then the third antecedent was reproduced; if the number returned was more than 0.647, but less than or equal to $0.647 + 0.353 = 1.0$, then the fourth antecedent was reproduced. On carrying out the operation of reproduction, the first, the fourth, the third, and again the fourth antecedents were reproduced (the second antecedent, which has the least fitness of all the antecedents, was not reproduced, and hence can be considered to have been deleted). The reproduced antecedents have been renumbered as shown in Fig. 4.3.

Antecedent number i	Antecedent	Fitness (F_i)
1	11 001 01	0.375
2	01 111 10	0.750
3	01 011 11	0.750
4	01 111 10	0.750

FIGURE 4.3 Antecedents obtained after carrying out reproduction on the antecedents of Fig. 4.2. The first antecedent given in Fig. 4.2 was reproduced as the first antecedent above, the fourth as the second above, the third as the third above, and the fourth antecedent was reproduced again as the fourth above. The fitness of each antecedent above was obtained from Fig. 4.2. Then the operation of mutation was carried out with a probability of 0.001, as explained in Section 4.4. The operation changed only the leftmost bit of the fourth antecedent above from 0 to 1. The antecedents after mutation are shown in Fig. 4.4.

Antecedent number (i)	Antecedent	Training patterns that the antecedent fits	Fitness (F_i)	Crossover Mate	Crossover Site
1	11 001 01	T1, T3, T6	0.375	2	3
2	01 111 10	T1, T2, T5, T6, T7, T8	0.750	1	3
3	01 011 11	T1, T2, T3, T6, T7, T8	0.750	4	6
4	11 111 10	T1, T2, T4, T5, T7, T8	0.750	3	6

FIGURE 4.4 Antecedents obtained after carrying out mutation (with a probability of 0.001) on the antecedents given in Fig. 4.3. The first three antecedents above are the same as in Fig. 4.3. The fourth antecedent above differs from the fourth antecedent of Fig. 4.3 in only the leftmost bit. The fitness of the first three antecedents was taken from Fig. 4.3, and the fitness of the fourth antecedent was calculated as in Fig. 4.2. Next, the operation of crossover is carried out. A random number generator chose antecedents 1 and 2 to be mates with the crossover site at 3, and antecedents 3 and 4 to be mates with the crossover site at 6. Having a crossover site within an attribute value helps produce a new value for that attribute. The antecedents obtained after carrying out crossover are shown in Fig. 4.5.

Antecedent number (i)	Antecedent	Training patterns that the antecedent fits	Fitness (F_i)
1	11 011 10	T1, T2, T3, T4, T5, T6, T7, T8	1.0
2	01 101 01	T3, T6, T8	0.375
3	01 011 10	T1, T2, T3, T5, T6, T7, T8	0.875
4	11 111 11	T2, T4, T7	0.375

FIGURE 4.5   Antecedents obtained after carrying out crossover on the antecedents given in Fig. 4.4. The first two antecedents above were obtained by carrying out crossover on the first two antecedents given in Fig. 4.4 at crossover site 3; and the third and fourth antecedents were obtained by carrying out crossover on the third and fourth antecedents given in Fig. 4.4 at crossover site 6. The fitness of each antecedent above was calculated as in Fig. 4.2. The fourth antecedent (a string of all 1's) denotes TRUE; it will cover every training pattern, but it should cover only training patterns T2, T4, and T7, which are the ones it, therefore, fits. The first antecedent fits all training patterns; its fitness becomes 1 and it can be used in a prule for recognizing professors.

If EATS = fried,
and FOOTWEAR = sandals,
then the pattern belongs to the class professor.

If EATS = roasted,
and FOOTWEAR = sandals,
then the pattern belongs to the class professor.

Coincidentally, the two prules happen to be identical to prules SG1 and SG2 (Fig. 1.3) obtained by the SpecToGen procedure.

The operations of reproduction, mutation, and crossover have been carried out once each in Figs 4.2 to 4.5. In practice, however, we may need to iterate much more, repeatedly carrying out these operations, hundreds—or perhaps thousands—of times.

As is clear from the description of the evolutionary procedure and the above example, with each successive generation, we try to produce antecedents that fit more and more training patterns. Nonetheless, there is no guarantee that the procedure will terminate with an antecedent having the highest possible fitness. There is arbitrariness in how and in what order we carry out the operations of reproduction, mutation, and crossover. It can be argued that the evolutionary procedure employs what mathematicians often call 'trial and error'. By viewing each bit in a binary string as a 'gene', some writers refer to the evolutionary procedure as a 'genetic algorithm'.

4.6 Modifications in the Evolutionary Procedure

The following are some of the modifications which can be adopted before implementing the evolutionary procedure.

1. The evolutionary procedure described in Section 4.5 develops antecedents to be used in prules for recognizing a given class C. If an antecedent covers only few training patterns of class C, then while carrying out mutation on the antecedent, change only a 0 bit to a 1 bit, never a 1 bit to a 0 bit. This will allow an attribute to have alternative values, hence making the antecedent more general, thereby perhaps covering more training patterns of class C (the definition of when an antecedent is more general than another is given in Section 1.4). As a complement to this, if the antecedent covers too many training patterns that are not in class C, then while carrying out mutation on the antecedent, change only a 1 bit to a 0 bit, never a 0 bit to a 1 bit. This will make the antecedent more specific, thereby perhaps covering fewer training patterns that are not in class C. Both these modifications are expected to increase the fitness of the said antecedent.

2. The crossover described in Section 4.4 is a *single-point crossover*, since there is a single crossover site. We can have more than one crossover sites. A *two-point crossover* carried out on two b-bit strings β and δ, where

    $$\beta = \beta_1 \; \beta_2 \; \beta_3$$
    $$\delta = \delta_1 \; \delta_2 \; \delta_3$$

    and where the length of β_i is equal to the length of δ_i, for $1 \leq i \leq 3$, produces the two strings

    $$\beta_1 \; \delta_2 \; \beta_3$$
    $$\delta_1 \; \beta_2 \; \delta_3$$

    For instance, from the strings

01101
10000

with crossover sites at 1 and 3, the two-point crossover produces the strings

00001
11100

A $(b-1)$-*point crossover* on b-bit strings β and δ, produces strings that take their bit alternately from β and δ. Thus the 4-point crossover on the strings

01101
10000

produces the strings

00101
11000

Given two strings, we can use a random number generator to decide at which sites (one or more) the crossover should be carried out on the two strings.

3. Suppose we mutate an antecedent α_1 to produce antecedent α_2. Of the two antecedents, we retain the one with the higher fitness and delete the other. Thus, it is not necessary that we delete α_1 and retain α_2.

4. Suppose we carry out crossover on antecedents α_1 and α_2 to produce antecedents α_3 and α_4. Of the four antecedents, we retain the two antecedents with the highest fitness, and delete the other two.

Summary

An evolutionary procedure starts with an initial population of antecedents that can be used in prules to recognize a given class. By applying the operations of reproduction, mutation, and crossover, we try to obtain antecedents that fit more training patterns compared to the initial antecedents. We may have to carry out these operations many times, before we obtain antecedents that meet our needs. The evolutionary procedure, however, is not guaranteed to produce such antecedents.

Exercises

1. Write a computer program in a language of your choice to implement the evolutionary procedure described in Section 4.5.

2. Figures 4.2 to 4.5 give an example on developing an antecedent for a prule to recognize professors of the professor–student training set given in Fig. 1.1. Use the program you wrote in the first exercise above for developing one or more antecedents to be used in prules for recognizing students. Generate the initial set of antecedents randomly. After how many generations did the program terminate? Next, generate the initial set of antecedents based on your intuition. After how many generations did the program terminate this time?

3. Use the program you wrote in the first exercise above to develop prules for the classification of computer science departments, from the training set described in the tenth exercise of Chapter 2. Report your results in classifying the patterns of the recall set.

4. The evolutionary procedure employs what mathematicians call the 'trial and error' method. Argue for or against it.

Learning Objectives

This chapter contains the following topics:

- prior, posterior, and conditional probabilities
- maximizing posterior probabilities using Bayes classification
- reducing computation assuming the class conditional independence of attributes in Naive Bayes classification
- maximum-likelihood and Bayesian estimates of probabilities
- estimating probabilities from the professor–student training set of the first chapter, and the Naive Bayes classifications of the patterns of the corresponding recall set
- classifying patterns by minimizing the risk of misclassification
- transforming a Naive Bayes classifier into a linear classifier by using binary attributes
- Bayes or Naive Bayes classifiers using probability density functions for continuous attribute values

5

Bayes Classification

5.1   Simplifying Bayes Classification

A Bayes classifier does not use prules to classify patterns. It estimates probabilities of occurrence of different attribute values for the different classes in a training set. It then uses these probabilities to classify recall patterns. Let

- $\bar{A}$ be an array of $M \geq 1$ attributes $A_1, A_2, \ldots, A_M$ for the patterns of a training set, and
- $P(\bar{A})$ be the probability that a training pattern has attribute array $\bar{A}$, regardless of the class to which the pattern belongs, the attributes having discrete values.

Suppose the training set has patterns from $m \geq 1$ classes $C_1, C_2, \ldots, C_m$. For $1 \leq k \leq m$, we define the following probabilities.

- $P(C_k)$ is the probability that a training pattern belongs to class C_k. It is also known as the *prior* probability of class C_k.
- $P(C_k|\bar{A})$ is the probability that a training pattern with attribute array $\bar{A}$ belongs to class C_k. This is also known as the *posterior* probability of C_k for a given $\bar{A}$. The attributes have discrete values.
- $P(\bar{A}|C_k)$ is the *conditional* probability that a training pattern of class C_k has attribute array $\bar{A}$, the attributes having discrete values.
- $p(\bar{A}|C_k)$ is the conditional probability density that a training pattern of class C_k has attribute array $\bar{A}$, the attributes having continuous values.

As is the convention, upper case P is being used for conditional probability when the attribute values are discrete, and lower case p is being used when the attribute values are continuous.

To classify a pattern with attribute array $\bar{A}$, we assign it to the class most probable for $\bar{A}$. This means we estimate the posterior probability $P(C_k|\bar{A})$ for each of the classes, that is, for $k = 1$ to m. We then assign the pattern to the class with the highest such probability (if two or more classes have equal highest posterior probabilities, then the pattern is rejected). It is briefly referred to as *maximizing* $P(C_k|\bar{A})$. According to the Bayes' theorem of probability theory

$$P(C_k|\bar{A}) = \frac{P(C_k)P(\bar{A}|C_k)}{P(\bar{A})}$$

Maximizing $P(C_k|\bar{A})$ is the same as maximizing the right-hand side of the above equation. The right-hand side denominator contains $P(\bar{A})$, which remains constant for all classes. So its value need not be estimated, hence maximizing $P(C_k|\bar{A})$ is the same as maximizing

$$P(C_k)P(\bar{A}|C_k) \qquad\qquad (5.1)$$

After they have been estimated from the training set, probabilities $P(C_k)$ and $P(\bar{A}|C_k)$ can be used to classify a pattern with attribute array $\bar{A}$. For most domains, estimating $P(\bar{A}|C_k)$ needs an impractically large training set, as we need to consider all possible values for the attributes A_1 to A_M. Moreover, it needs a lot of computation, which, in turn, can require a lot of time. So it is usually assumed that the attributes A_1 to A_M are class conditionally independent, which means it is often assumed that

$$P(\bar{A}|C_k) = \prod_{i=1}^{M} P(A_i|C_k)$$

After making the above assumption, the classifier is called the *Naive* Bayes classifier, in which to classify a pattern with attributes A_1 to A_M, we maximize

$$P(C_k)\prod_{i=1}^{M} P(A_i|C_k) \qquad\qquad (5.2)$$

which is obtained by substituting $\prod_{i=1}^{M} P(A_i|C_k)$ for $P(\bar{A}|C_k)$ in Eqn (5.1).

5.2 Estimation of Probabilities

Suppose that the number of patterns in class C_k is $|C_k|$, for $1 \leq k \leq m$. Then under the conventional estimation, known as the *maximum-likelihood* estimation,

$$P(C_k) = \frac{|C_k|}{\sum_{j=1}^{m} |C_j|} \qquad\qquad (5.3)$$

Under an alternative estimation, called the *Bayesian* estimation,

$$P(C_k) = \frac{|C_k| + 1}{m + \sum\limits_{j=1}^{m} |C_j|} \tag{5.4}$$

In the Bayesian estimation, we assume the number of patterns of each class to be one more than actually present in the training set. For a typical training set with a lot of patterns (that is, $\sum_{j=1}^{m} |C_j|$ is a large number), both the maximum-likelihood and Bayesian estimates give approximately equal values. Moreover, for both estimates, $\sum_{k=1}^{m} P(C_k) = 1$, as it should be. One difference exists between the two estimations: in the maximum-likelihood estimation, $P(C_k) \geq 0$; in the Bayesian estimation, because the numerator of the above equation is always one or more, $P(C_k) > 0$. In other words, in the Bayesian estimation a probability is never equal to zero. This is useful in practice, as we shall soon see.

As an example, the professor–student training set given in Fig. 1.1 has 3 patterns belonging to the class professor, and 5 to the class student. There being two classes, the value of m is 2. Hence, from Eqn (5.3), the maximum-likelihood estimate of $P(\text{professor})$ is $3/(3+5)$, that is $3/8$; and the Bayesian estimate of $P(\text{professor})$, from Eqn (5.4), is $(3+1)/(2+3+5)$, that is $4/10$, which may be simplified to $2/5$.

To maximize $P(C_k) \prod_{i=1}^{M} P(A_i|C_k)$ of Eqn (5.2), the other probability estimates required are those for $\prod_{i=1}^{M} P(A_i|C_k)$. Let

- the possible values for attribute A_i be $v_{i_1}, v_{i_2}, \ldots, v_{i_n}$, for $1 \leq i \leq M$ and
- $|C_k^{ij}|$ be the number of training patterns of class C_k for which the value of attribute A_i is v_j, for $1 \leq k \leq m$, for $1 \leq i \leq M$, and for $i_1 \leq j \leq i_n$.

Then, for $1 \leq k \leq m$, for $1 \leq i \leq M$, and for $i_1 \leq j \leq i_n$, according to the maximum-likelihood estimation

$$P[(A_i = v_j)|C_k] = \frac{|C_k^{ij}|}{|C_k|} \tag{5.5}$$

and according to the Bayesian estimation

$$P[(A_i = v_j)|C_k] = \frac{|C_k^{ij}| + 1}{i_n + |C_k|} \tag{5.6}$$

In the Bayesian estimation, we assume that the number of training patterns of class C_k for which the value of attribute A_i is v_j is one more than actually present. Owing to this, as mentioned above, in the Bayesian estimation, no probability is ever equal to zero.

To illustrate, attribute EATS of the training set given in Fig. 1.1 has 3 possible values (baked, roasted, and fried). Of the 5 student patterns, none has the value of EATS = roasted. Therefore, from Eqn (5.5), the maximum-likelihood estimate of $P[(\text{EATS} = \text{roasted}) \mid \text{student}]$ is 0/5, and the Bayesian estimate, from Eqn (5.6), is $(0+1)/(3+5)$, which becomes 1/8.

The maximum-likelihood and Bayesian estimates of the various probabilities from the professor–student training set given in Fig. 1.1 are shown in Fig. 5.1. These are the probability estimates said to have been learned by the classifier. The correctness of the probability estimates given in Fig. 5.1 can be checked using Eqns (5.5) and (5.6). The fractions given in the figure have not been simplified, to avoid any confusion; for instance, a probability estimate of 4/10 appears as such, and has not been simplified to 2/5. Since our training set is not large, the maximum-likelihood estimation of a probability is not as close to the Bayesian estimation as it would have been, if the training set had been large. Nevertheless, the usefulness of Bayesian estimation will become clear in the next Section 5.3.

5.3 Classifying Professor–Student Patterns

Let us classify pattern R3 of the professor–student recall set given in Fig. 1.2. The attribute values for the pattern are HABIT = gabby, EATS = roasted, and FOOTWEAR = clogs. Using the maximum-likelihood estimates from Fig. 5.1 and the Naive Bayes classifier of Eqn (5.2), the value of

$P(\text{professor}) \times P[(\text{HABIT} = \text{gabby}) \mid \text{professor}]$

$\times P[(\text{EATS} = \text{roasted}) \mid \text{professor})] \times P[(\text{FOOTWEAR} = \text{clogs}) \mid \text{professor}]$

is $3/8 \times 2/3 \times 1/3 \times 0/3 = 0$, and the value of

$$P(\text{student}) \times P[(\text{HABIT} = \text{gabby}) \,|\, \text{student}]$$

$$\times P[(\text{EATS} = \text{roasted}) \,|\, \text{student}] \times P[(\text{FOOTWEAR} = \text{clogs}) \,|\, \text{student}]$$

	Estimates	
Probability	Maximum-likelihood	Bayesian
$P(\text{professor})$	3/8	4/10
$P[(\text{HABIT} = \text{gabby}) \,|\, \text{professor}]$	2/3	3/5
$P[(\text{HABIT} = \text{quiet}) \,|\, \text{professor}]$	1/3	2/5
$P[(\text{EATS} = \text{baked}) \,|\, \text{professor}]$	0/3	1/6
$P[(\text{EATS} = \text{fried}) \,|\, \text{professor}]$	2/3	3/6
$P[(\text{EATS} = \text{roasted}) \,|\, \text{professor}]$	1/3	2/6
$P[(\text{FOOTWEAR} = \text{clogs}) \,|\, \text{professor}]$	0/3	1/5
$P[(\text{FOOTWEAR} = \text{sandals}) \,|\, \text{professor}]$	3/3	4/5
$P(\text{student})$	5/8	6/10
$P[(\text{HABIT} = \text{gabby}) \,|\, \text{student}]$	3/5	4/7
$P[(\text{HABIT} = \text{quiet}) \,|\, \text{student}]$	2/5	3/7
$P[(\text{EATS} = \text{baked}) \,|\, \text{student}]$	3/5	4/8
$P[(\text{EATS} = \text{fried}) \,|\, \text{student}]$	2/5	3/8
$P[(\text{EATS} = \text{roasted}) \,|\, \text{student}]$	0/5	1/8
$P[(\text{FOOTWEAR} = \text{clogs}) \,|\, \text{student}]$	3/5	4/7
$P[(\text{FOOTWEAR} = \text{sandals}) \,|\, \text{student}]$	2/5	3/7

FIGURE 5.1 Maximum-likelihood and Bayesian estimates of prior and conditional probabilities from the professor–student training set given in Fig. 1.1. A few of the maximum-likelihood estimates are zero but none of the Bayesian estimates are so. The maximum-likelihood estimates of the prior probabilities $P(\text{professor})$ and $P(\text{student})$ are obtained from Eqn (5.3), and their Bayesian estimates from Eqn (5.4). The remaining are conditional probabilities: their maximum-likelihood estimates are obtained from Eqn (5.5), and the Bayesian estimates from Eqn (5.6).

is $5/8 \times 3/5 \times 0/5 \times 3/5 = 0$. The pattern is rejected because the values for both the classes are zero. For each class, one zero probability has nullified the influence of the other probabilities. This is a disadvantage of using the maximum-likelihood estimates of probabilities. Instead, using the Bayesian estimates from Fig. 5.1, the value of

$P(\text{professor}) \times P[(\text{HABIT} = \text{gabby}) \mid \text{professor}]$

$\times P[(\text{EATS} = \text{roasted}) \mid \text{professor}] \times P[(\text{FOOTWEAR} = \text{clogs}) \mid \text{professor}]$

is $2/5 \times 3/5 \times 1/3 \times 1/5 = 0.016$, and the value of

$P(\text{student}) \times P[(\text{HABIT} = \text{gabby}) \mid \text{student}]$

$\times P[(\text{EATS} = \text{roasted}) \mid \text{student}] \times P[(\text{FOOTWEAR} = \text{clogs}) \mid \text{student}]$

is $3/5 \times 4/7 \times 1/8 \times 4/7 = 0.0245$. Since the value 0.0245 obtained for student is more than 0.016 obtained for professor, recall pattern R3 is classified as student.

Thus, it is observed that pattern R3 was rejected when maximum-likelihood estimation was used, but was classified when Bayesian estimation was used. Bayesian estimates have accordingly been found to be useful in practice.

Using the Bayesian estimates of probabilities from Fig. 5.1, the Naive Bayes classifications of the four professor–student recall patterns given in Fig. 1.2 are shown in Fig. 5.2. The classifications are incidentally identical to those shown in Fig. 2.11 for the decision tree of Fig. 2.9 (the tree obtained by maximizing the ratio of information gain).

NAME of recall pattern	Attributes			CLASSIFICATION
	HABIT	**EATS**	**FOOTWEAR**	
R1	Quiet	Baked	Clogs	Student
R2	Quiet	Roasted	Sandals	Professor
R3	Gabby	Roasted	Clogs	Student
R4	Quiet	Roasted	Clogs	Student

FIGURE 5.2   Classifications of the professor–student recall patterns given in Fig. 1.2, using the Bayesian estimates of probabilities in the Naive Bayes classifier. The Bayesian estimates were taken from Fig. 5.1, and the Naive Bayes classifier of Eqn (5.2) was used.

5.4 Minimizing Risk

Some misclassifications can cause more loss than others: classifying a cancerous tumour as benign may lead to premature death; classifying a benign tumour as cancerous, to unnecessary therapy. The former is considered to cause more loss than the latter.

Let us assume there is no loss when a pattern is classified correctly, but there is a loss when the pattern is misclassified. For the m classes C_1 to C_m, let $\lambda(C_i)$ be the loss in misclassifying a pattern of class C_i, where $1 \leq i \leq m$. This means we are assuming that the loss is the same regardless of which class the pattern is misclassified into. Suppose there is a pattern with attribute array $\bar{A}$, and we need to classify it. The expected loss (known as the *risk*) in classifying the pattern into class C_k is

$$\sum_{i \neq k}^{m} \lambda(C_i) P(C_i | \bar{A})$$

There will be no loss if the pattern actually belongs to class C_k, hence we have $i \neq k$ with the summation sign. The above equation can be rewritten as

$$\sum_{i=1}^{m} \lambda(C_i) P(C_i | \bar{A}) - \lambda(C_k) P(C_k | \bar{A})$$

We want to minimize the risk in classifying, therefore we minimize the above equation. The first term of the equation is a constant. Accordingly, we minimize the above equation by maximizing

$$\lambda(C_k) P(C_k | \bar{A})$$

In other words, to minimize the risk, we classify the pattern into that class C_k for which the above equation has the highest value. Following the same reasoning as that used to derive Eqn (5.2), we can say that, to minimize risk in the Naive Bayes classifier, we maximize

$$\lambda(C_k) P(C_k) \prod_{i=1}^{M} P(A_i | C_k) \tag{5.7}$$

In Section 5.3, we have not considered risk when we classify recall pattern R3 of the professor–student training set given in Fig. 1.2 by maximizing Eqn (5.2).

Suppose we need to reclassify the pattern by minimizing risk, where $\lambda(\text{professor}) = 2$ and $\lambda(\text{student}) = 1$. In other words, the loss in misclassifying a professor is twice the loss in misclassifying a student. The attribute values for the pattern are HABIT = gabby, EATS = roasted, and FOOTWEAR = clogs. Using the Bayesian probability estimates from Fig. 5.1, and the Naive Bayes classifier of Eqn (5.7), the value of

$\lambda(\text{professor}) \times P(\text{professor}) \times P[(\text{HABIT} = \text{gabby}) \,|\, \text{professor}]$

$\times P[(\text{EATS} = \text{roasted}) \,|\, \text{professor}] \times P[(\text{FOOTWEAR} = \text{clogs}) \,|\, \text{professor}]$

is $2 \times 2/5 \times 3/5 \times 1/3 \times 1/5 = 0.032$, and the value of

$\lambda(\text{student}) \times P(\text{student}) \times P[(\text{HABIT} = \text{gabby}) \,|\, \text{student}]$

$\times P[(\text{EATS} = \text{roasted}) \,|\, \text{student}] \times P[(\text{FOOTWEAR} = \text{clogs}) \,|\, \text{student}]$

is $1 \times 3/5 \times 4/7 \times 1/8 \times 4/7 = 0.0245$.

Since the value 0.032 obtained for professor is more than the 0.0245 obtained for student, recall pattern R3 is classified as professor. This contrasts with the classification in Section 5.3, where, by not considering risk, recall pattern R3 is classified as student.

For the rest of the discussion in this chapter, it will be assumed that $\lambda(C_k) = 1$, for $1 \le k \le m$, that is, the loss in any misclassification is 1. Under this assumption, Eqn (5.7) simplifies to Eqn (5.2). It can be said that Eqn (5.2) is a special case of Eqn (5.7).

5.5 Naive Bayes with Binary Attributes

A *binary* attribute has two possible values; for example, attribute HABIT in the professor–student patterns given in Fig. 1.1 has two possible values: gabby and quiet. We could as well have had HABIT's possible values being 0 and 1. In general, any binary attribute can be coded so that the possible values are 0 and 1.

Consider a training set of patterns whose every attribute A_1 to A_M is binary, each attribute's possible values being 0 and 1. For the m classes C_1 to C_m, suppose we have estimated from the training set the prior probability $P(C_k)$ and the conditional probability $P(A_i|C_k)$, for $1 \leq k \leq m$ and $1 \leq i \leq M$, as described in Section 5.2. Let us define

$$a_{ki} = P[(A_i = 1)|C_k] \tag{5.8}$$

Since the binary attribute A_i can have a value either 0 or 1, we will have $P[(A_i = 1)|C_k] + P[(A_i = 0)|C_k] = 1$. Therefore, $P[(A_i = 0)|C_k] = 1 - P[(A_i = 1)|C_k]$. Accordingly, from Eqn (5.8), we can write

$$P[(A_i = 0)|C_k] = 1 - a_{ki} \tag{5.9}$$

To classify a pattern with attribute array $\bar{A}$, by the Naive Bayes classifier, we will be maximizing Eqn (5.2), that is, we will be maximizing

$$P(C_k) \prod_{i=1}^{M} P(A_i|C_k)$$

Since logs are monotonic, we can alternatively maximize

$$\log P(C_k) + \sum_{i=1}^{M} \log P(A_i|C_k)$$

We will evaluate the above equation for classes $C_1, C_2, \ldots, C_m$ to find out for which class its value is the greatest, so that we can classify the pattern in that class. Let us define a function $g_k(\bar{A})$ such that

$$g_k(\bar{A}) = \log P(C_k) + \sum_{i=1}^{M} \log P(A_i|C_k) \tag{5.10}$$

The second term in the above equation

$$\log P(A_i|C_k) = A_i \log a_{ki} + (1 - A_i) \log(1 - a_{ki})$$
identity holds for possible values 0 and 1 of A_i
with a_{ki} as defined in Eqns (5.8) and (5.9)

$$= \log(1 - a_{ki}) + A_i[\log a_{ki} - \log(1 - a_{ki})]$$

$$= \log(1 - a_{ki}) + A_i \log \frac{a_{ki}}{1 - a_{ki}}$$

Substituting the right-hand side of the above equation for $\log P(A_i|C_k)$ in Eqn (5.10), we get

$$g_k(\bar{A}) = \log P(C_k) + \sum_{i=1}^{M} \log(1 - a_{ki}) + \sum_{i=1}^{M} A_i \log \frac{a_{ki}}{1 - a_{ki}} \qquad (5.11)$$

This is a linear equation in A_i. The equation can be viewed as

$$g_k(\bar{A}) = w_{k0} + \sum_{i=1}^{M} w_{ki} A_i$$

where

$$w_{k0} = \log P(C_k) + \sum_{i=1}^{M} \log(1 - a_{ki})$$

and

$$w_{ki} = \log \frac{a_{ki}}{1 - a_{ki}}$$

for $0 < a_{ki} < 1$, such that $1 \leq k \leq m$ and $1 \leq i \leq M$. The w's are viewed as *weights*, and $g_k(\bar{A})$ is the linear *discriminant function* for class C_k. Given the attribute array $\bar{A}$ of a pattern to be classified, the value of the discriminant function for each class is calculated, and then the pattern is assigned to the class with the discriminant function having the largest value. If each pattern is represented as a point in the M-dimensional A_1-A_2- $\cdots$ -A_M coordinate space, then each linear discriminant function represents a hyperplane that separates one class from another: the classes are *linearly separable*. A classifier that uses linear discriminant functions is said to be a *linear* classifier. Accordingly, a Naive Bayes classifier applied to patterns with binary attributes can be formulated as a linear classifier (more on linear classifiers is given in Sections 6.5, 7.2, and 7.3, and in Chapter 8).

Consider an example of three classes C_1, C_2, and C_3. The attribute array $\bar{A}$ contains a single binary attribute A_1. Therefore, $M = 1$. From a training set, suppose we have estimated the following probabilities:

$$P(C_1) = 0.3$$
$$P[(A_1 = 1)|C_1] = 0.8 = a_{11}$$
$$P[(A_1 = 0)|C_1] = 0.2 = 1 - a_{11}$$

$$P(C_2) = 0.5$$
$$P[(A_1 = 1)|C_2] = 0.4 = a_{21}$$
$$P[(A_1 = 0)|C_2] = 0.6 = 1 - a_{21}$$

$$P(C_3) = 0.2$$
$$P[(A_1 = 1) \mid C_3] = 0.3 = a_{31}$$
$$P[(A_1 = 0) \mid C_3] = 0.7 = 1 - a_{31}$$

Let us classify a pattern whose attribute $A_1 = 1$. We need to evaluate the discriminant function as given in Eqn (5.11) for each of the three classes.

$$g_1(\bar{A}) = \log P(C_1) + \log a_{11} + A_1 \log \frac{a_{11}}{1 - a_{11}} = -0.6198$$

$$g_2(\bar{A}) = \log P(C_2) + \log a_{21} + A_1 \log \frac{a_{21}}{1 - a_{21}} = -0.6989$$

$$g_3(\bar{A}) = \log P(C_3) + \log a_{31} + A_1 \log \frac{a_{31}}{1 - a_{31}} = -1.2219$$

Since the value of $g_1(\bar{A})$ is the largest, the pattern with $A_1 = 1$ is classified into class C_1. Now let us classify a pattern whose attribute $A_1 = 0$. We need to evaluate the discriminant function for each of the three classes.

$$g_1(\bar{A}) = \log P(C_1) + \log a_{11} + A_1 \log \frac{a_{11}}{1 - a_{11}} = -1.2219$$

$$g_2(\bar{A}) = \log P(C_2) + \log a_{21} + A_1 \log \frac{a_{21}}{1 - a_{21}} = -0.5228$$

$$g_3(\bar{A}) = \log P(C_3) + \log a_{31} + A_1 \log \frac{a_{31}}{1 - a_{31}} = -0.8539$$

Since the value of $g_2(\bar{A})$ is the largest, the pattern with $A_1 = 0$ is classified into class C_2.

Let us now use Eqn (5.2) of the Naive Bayes classifier and do the same two classifications to check whether the results are identical. For the pattern whose attribute $A_1 = 1$:

$$P(C_1) \times P[(A_1 = 1)|C_1] = 0.3 \times 0.8 = 0.24$$

$$P(C_2) \times P[(A_1 = 1)|C_2] = 0.5 \times 0.4 = 0.20$$

$$P(C_3) \times P[(A_1 = 1)|C_3] = 0.2 \times 0.3 = 0.06$$

Since the value 0.24 obtained for class C_1 is the largest, the pattern is classified into class C_1. This is the same as the classification done above by the linear classifier. Now let us classify the pattern whose attribute $A_1 = 0$:

$$P(C_1) \times P[(A_1 = 0)|C_1] = 0.3 \times 0.2 = 0.06$$

$$P(C_2) \times P[(A_1 = 0)|C_2] = 0.5 \times 0.6 = 0.30$$

$$P(C_3) \times P[(A_1 = 0)|C_3] = 0.2 \times 0.7 = 0.14$$

Since the value 0.30 obtained for class C_2 is the largest, the pattern is classified into class C_2. This is the same as the classification done above by the linear classifier. So for both classifications, Eqn (5.2) and the linear classifier give identical results.

5.6   Continuous Attribute Values

Readers may be familiar with the terms *mean, variance, standard deviation,* and *covariance*. Nonetheless, Fig. 5.3 reviews these terms briefly.

In Section 5.2, we learnt how to estimate conditional probabilities when attribute values are discrete. Attribute values can, however, be continuous, for example, the height and weight of a person. In this section, we will discuss how to estimate conditional probabilities when attribute values are continuous.

Adapting Eqn (5.2) to Naive Bayes with continuous attribute values, we maximize $P(C_k) \prod_{i=1}^{M} p(A_i|C_k)$ for $1 \leq k \leq m$, the classes being C_1 to C_m. We need to obtain, from the training set, the distribution of attribute A_i over class C_k. In practice, we often assume the attribute to have a Normal

---

Let $\bar{x}$ be an array of $n \geq 1$ numbers $x_1, x_2, \ldots, x_n$. Then the *mean* μ_x of the array is defined to be

$$\frac{1}{n}\sum_{i=1}^{n} x_i$$

The *variance* σ^2 of array $\bar{x}$ is defined to be

$$\frac{1}{n}\sum_{i=1}^{n}(x_i - \mu_x)^2$$

The positive square root of the variance is defined to be the *standard deviation* σ of the array.

Let there be another array $\bar{y}$ of n numbers $y_1, y_2, \ldots, y_n$ numbers, and let their mean be μ_y. The *covariance* σ_{xy} of arrays $\bar{x}$ and $\bar{y}$ is defined to be

$$\frac{1}{n}\sum_{i=1}^{n}(x_i - \mu_x)(y_i - \mu_y)$$

FIGURE 5.3   A review of mean, variance, standard deviation, and covariance.

(or Gaussian) distribution, hence for $1 \leq i \leq M$, we have

$$p(A_i|C_k) = \frac{1}{\sigma_i\sqrt{2\pi}}e^{-(A_i-\mu_i)^2/2\sigma_i^2}$$

where μ_i is the mean, σ_i is the standard deviation, and σ_i^2 is the variance of attribute A_i over the training patterns of class C_k.

If we do not want to assume class conditional independence of the attributes, we use the Bayes classifier, not Naive Bayes. Then, adapting Eqn (5.1) to continuous attribute values, we maximize $P(C_k)p(\bar{A}|C_k)$. For that, we need, from the training set, the distribution of the attribute array $\bar{A} = A_1, A_2, \ldots, A_M$ over class C_k. In practice, we often assume a multivariate normal distribution, hence we have

$$p(\bar{A}|C_k) = \frac{1}{\sqrt{|\Sigma|(2\pi)^M}}e^{[-(\bar{A}-\bar{\mu})^t\Sigma^{-1}(\bar{A}-\bar{\mu})]/2}$$

in which

$$
\Sigma = \begin{bmatrix}
\sigma_1^2 & \sigma_{12} & \sigma_{13} & \cdots & \sigma_{1M} \\
\sigma_{21} & \sigma_2^2 & \sigma_{23} & \cdots & \sigma_{2M} \\
\sigma_{31} & \sigma_{32} & \sigma_3^2 & \cdots & \sigma_{3M} \\
\vdots & \vdots & \vdots & \ddots & \vdots \\
\sigma_{M1} & \sigma_{M2} & \sigma_{M3} & \cdots & \sigma_M^2
\end{bmatrix}
$$

and

- $|\Sigma|$ is the determinant of Σ.
- σ_i^2 is the variance of A_i over the training patterns of class C_k for $1 \le i \le M$.
- σ_{ij} is the covariance of A_i and A_j over the training patterns of class C_k, for $1 \le i \le M$ and $1 \le j \le M$, where $i \neq j$.
- $\bar{\mu} = \bar{\mu}_1, \bar{\mu}_2, \ldots, \bar{\mu}_M$, where $\bar{\mu}_i$ is the mean of A_i over the training patterns of class C_k.
- $(\bar{A} - \bar{\mu})^t$ is the transpose of $(\bar{A} - \bar{\mu})$.

5.7   Performance of Bayes Classifier

The Bayes classifier makes the optimal decision in assigning a pattern to the most probable class. In theory, a Bayes classifier performs the least misclassifications as compared to any other classifier. In practice, the Bayes classifier usually does well, but not so well as the theory says it should. This happens mainly because of inaccuracies in estimating the probabilities. Moreover, if the Naive Bayes classifier is used, then the assumption of the attributes being class conditionally independent may not be true.

Summary

Given the attribute values of a pattern, a Bayes classifier recognizes the pattern by maximizing the posterior probability over the classes. To reduce computation, the Naive Bayes classifier, in which the attributes are assumed to be class conditionally independent, is often used. The Bayesian estimation of probabilities is at times more useful in classification than the maximum-likelihood estimation. The Bayesian classifier can alternatively base its classification on minimizing the risk of misclassification. With binary attributes, the Naive Bayes classifier becomes a linear classifier. When the attribute values are continuous, the probability density functions can be used to estimate the conditional probabilities.

Exercises

1. Manually verify the probability estimates given in Fig. 5.1.

2. Obtaining the Bayesian estimates of probabilities from Fig. 5.1, manually carry out the calculations required to classify patterns R1, R2, and R4 in Fig. 5.2 by the Naive Bayes classifier using Eqn (5.2). The calculations required to classify pattern R3 are already given in the example of Section 5.3. Again, classify these patterns using the maximum-likelihood estimates of probabilities. Given $\lambda(\text{professor}) = 2$ and $\lambda(\text{student}) = 1$, reclassify the patterns by the Naive Bayes classifier minimizing risk, by using Eqn (5.7), and by using the Bayesian estimates of probabilities.

3. Obtaining the Bayesian estimates of probabilities from Fig. 5.1, manually show that the Naive Bayes classifier using Eqn (5.2) will correctly classify all the professor–student training patterns T1 to T8 given in Fig. 1.1.

4. Suppose attribute HABIT does not exist in the professor–student training set given in Fig. 1.1 and the recall set given in Fig. 1.2. After training manually on the eight training patterns, classify the four recall patterns by the Naive Bayes classifier [Eqn (5.2)], using Bayesian probability estimates. Are the classifications same as those in Fig. 5.2, in which HABIT is an attribute? If yes, why is it so? (*Hint*: See Fig. 2.9)

5. In a programming language of your choice, write a program to implement a Naive Bayes classifier using Eqn (5.7). In the training phase, it estimates the required prior and conditional probabilities from a given training set. The user can choose to have the program give either maximum-likelihood or Bayesian estimates of probabilities. In the testing phase, (a) the user provides the value $\lambda(C_i)$, the loss in misclassifying a pattern of class C_i, for each of the m classes, that

is, for $1 \leq i \leq m$; (b) the program uses the probability estimates it had made in the training phase to classify a given training or recall pattern. Use this program to repeat the first four exercises. It is shown in Section 5.4 that when the loss of every misclassification is 1, Eqn (5.7) becomes Eqn (5.2).

6. Use the program you wrote in the fifth exercise above to train a Naive Bayes classifier on the training set of computer science departments described in the tenth exercise of Chapter 2. Report your results of classifying the patterns of the recall set.

7. Extend the program you wrote in the fifth exercise above so that it can be used as a linear classifier when all attributes are binary.

8. In Section 3.1, you read how decision trees are adapted to handle missing attribute values. Can the Bayes procedure be adapted to handle missing attribute values? If not, why? If yes, describe the adaptations required. Remember, there may be missing attribute values in the training set. So, changes may be required in the estimation of probabilities from the training set. When there are missing attribute values in a recall pattern, changes may be required in the classification of the pattern.

Learning Objectives

This chapter contains the following topics:

- basis of nearest neighbour classification
- Euclid or city-block distance between patterns that have numeric attributes
- Hamming distance between patterns that have non-numeric attributes
- applying nearest neighbour classification to the professor–student recall patterns of the first chapter
- representing a class by its prototype
- the transformation of the assignment of a pattern to the class of the nearest prototype into a linear classifier
- relationship between the error rates of the nearest neighbour and Bayes classifiers
- modifications in the nearest neighbour classifier

6

Nearest Neighbour Classification

6.1 Underlying Idea

'Birds of a feather flock together' or 'You are like the company you keep' are popular sayings. A person living in a rich neighbourhood is classified as rich, whereas a person living in a slum is classified as poor. Of course, there can be exceptions. A rich person may live in a slum. In conversation, that person may be called an eccentric or an oddball; in statistics, the person will be called an outlier. An *outlier* is a pattern that is in some way unlike the other patterns of its class: in other words, it lies outside its class. Outliers are rare. This idea that most of the time our class is the same as that of our neighbours has been applied to a procedure called the *nearest neighbour* classification.

6.2 Numeric Attribute Values

To understand nearest neighbour classification using numeric attribute values, let us suppose the following.

- A_1, A_2, ... , A_M are the $M \geq 1$ numeric attributes of the patterns in a training set
- A_1', A_2', ... , A_M' are the values of the attributes for some training pattern, such that the value of A_i is A_i', for $1 \leq i \leq M$.
- A_1'', A_2'', ... , A_M'' are the values of the attributes for some recall pattern, such that the value of A_i is A_i'', for $1 \leq i \leq M$.

The above training pattern can be represented as a point in the M-dimensional A_1-A_2-$\cdots$-A_M coordinate space, such that the coordinate of A_i is A_i', for $1 \leq i \leq M$. Similarly, the above recall pattern can be represented as a point in this space, with the coordinate of A_i being A_i'', for $1 \leq i \leq M$. Then, the *Euclidean distance* between the training pattern and the recall pattern is defined to be

$$\sqrt{\sum_{i=1}^{M}(A_i'' - A_i')^2} \tag{6.1}$$

Each training pattern is viewed to be a *neighbour* of the recall pattern. The smaller the Euclidean distance between the recall pattern and a neighbour, the nearer the recall pattern is to that neighbour. To classify a given recall pattern by the nearest neighbour classifier, do the following.

1. Calculate the Euclidean distance between the recall pattern and each of its neighbours.
2. Assign the recall pattern to the class of its nearest neighbour.

If the nearest neighbour happens to be an outlier, then the recall pattern is likely to be misclassified. A modification to the procedure is to assign the recall pattern to the class most frequent among its $k \geq 1$ nearest neighbours, where the value of k is a design choice that one can make. In practice, one may try out different values of k and select the value that gives the best results. To obtain a value of k heuristically, often found useful in practice, start trying with an integer approximation of the square root of the number of training patterns in the class that has the fewest patterns.

An alternative to using the Euclidean distance is the city-block distance. The *city-block* (or *Manhattan*), *distance* between the two patterns of Eqn (6.1) is defined to be

$$\sum_{i=1}^{M} |A_i'' - A_i'| \tag{6.2}$$

In practice, the Euclidean distance is used more often than the city-block distance.

For an informal understanding of the Euclidean and city-block distances, think of a city in which the streets have been laid out in a rectangular grid in an M-dimensional space, and the two patterns are two points in the city. Then, the Euclidean distance is the straight-line distance you would fly from one pattern to the other, and the city-block distance is the distance you would walk along the streets from one pattern to the other.

A generalization of Euclidean and city-block distances is the *Minkowski* distance

$$\sum_{i=1}^{M} (|A_i'' - A_i'|^r)^{1/r}$$

For $r = 1$, the Minkowski distance becomes the city-block distance; for $r = 2$, it becomes the Euclidean distance.

In calculating Euclidean or city-block distances, we can choose to put different weights on the various attributes. Then, the weighted Euclidean distance can be obtained from the equation

$$\sqrt{\sum_{i=1}^{M} w_i (A_i'' - A_i')^2} \qquad (6.3)$$

and the weighted city-block distance can be obtained from the equation

$$\sum_{i=1}^{M} w_i |A_i'' - A_i'| \qquad (6.4)$$

where $w_i \geq 0$ is the weight put on attribute A_i, for $1 \leq i \leq M$. The higher the weight put on an attribute, the more important the attribute is considered to be. For example, to decide the class of therapy for a cancer patient, a doctor may consider the size of the patient's tumour to be more important than his age.

Equation (6.1) is a special case of Eqn (6.3), and Eqn (6.2) is a special case of Eqn (6.4), when $w_i = 1$, for $1 \leq i \leq M$. To save computation while using Eqn (6.1), the square root can be omitted because it is monotonic; it will not affect the classification results.

6.3 Non-numeric Attribute Values

If the attribute values are non-numeric, as they are for the professor–student patterns given in Figs 1.1 and 1.2, then the Euclidean or city-block distances cannot be calculated. Instead, the dissimilarity of a recall pattern from a given training pattern will need to be measured. This measure can depend on the domain of the patterns. A domain-independent measure, sometimes adopted, is the Hamming distance between the recall pattern and the training pattern. The *Hamming distance* between two patterns is defined to be the number of attributes for which the two patterns have different values. For example, the Hamming distance between the recall pattern R1 given in Fig. 1.2 having attributes values

HABIT = quiet, EATS = baked, FOOTWEAR = clogs

and the training pattern T1 given in Fig. 1.1 having attribute values

HABIT = gabby, EATS = baked, FOOTWEAR = clogs

is 1, because only the value of HABIT is different for the two patterns. The smaller the Hamming distance between a recall pattern and a training pattern, the nearer the two patterns are. To classify a given recall pattern:

1. Calculate the Hamming distance between the recall pattern and each of the training patterns.
2. Assign the recall pattern to the class of the training pattern (neighbour) nearest to it.

Figure 6.1 illustrates the classifications of the four professor–student recall patterns given in Fig. 1.2. The classifications in Fig. 6.1 are incidentally identical to the Naive Bayes classifications given in Fig. 5.2. Moreover, the classifications are also identical to the classifications in Fig. 2.11 for the decision tree shown in Fig. 2.9 (the tree obtained by maximizing the ratio of information gain).

As it is for numeric attributes, a modification to the above procedure is to assign the recall pattern to the class most frequent among its $k \geq 1$ nearest neighbours, where the value of k can be chosen as discussed in Section 6.2.

In calculating Hamming distance, we can choose to put different weights on the various attributes. Consider, recall pattern R1 given in Fig. 1.2 having attribute values

HABIT = quiet, EATS = baked, FOOTWEAR = clogs

and training pattern T7 given in Fig. 1.1 having attribute values

HABIT = gabby, EATS = fried, FOOTWEAR = sandals

The value of each of the three attributes is different for the two patterns. If we put a weight of 1 on HABIT, 3 on EATS, and 2 on FOOTWEAR, then the weighted Hamming distance between the two patterns is $1 + 3 + 2 = 6$. The weight of each attribute is 1 in the example shown in Fig. 6.1.

When attribute values are non-numeric, then the above procedure is also known as 'case-based reasoning'. At times, humans do case-based

Recall pattern	Training pattern	Attributes whose values differ in recall and training patterns	Hamming distance	Class of nearest training pattern	Classi-fication of recall pattern
R1	T1	H	1	S	S
	T2	H, E, F	3		
	T3	H, F	2		
	T4	E, F	2		
	T5	H, E	2		
	T6	F	1	S	
	T7	H, E, F	3		
	T8	E	1	S	
R2	T1	H, E, F	3		P
	T2	H	1	P	
	T3	H, E	2		
	T4	E	1	P	
	T5	H, E, F	3		
	T6	E	1	S	
	T7	H, E	2		
	T8	E, F	2		
R3	T1	E	1	S	S
	T2	F	1	P	
	T3	E, F	2		
	T4	H, E, F	3		
	T5	E	1	S	
	T6	H, E, F	3		
	T7	E, F	2		
	T8	H, E	2		
R4	T1	H, E	2		S
	T2	H, F	2		
	T3	H, E, F	3		
	T4	E, F	2		
	T5	H, E	2		
	T6	E, F	2		
	T7	H, E, F	3		
	T8	E	1	S	

FIGURE 6.1   Classifications of the four professor–student recall patterns given in Fig. 1.2 by assigning a recall pattern to the class of its nearest training pattern (see Fig. 1.1). In the third column from the left, H, E, and F are respectively HABIT, EATS, and FOOTWEAR. In the rightmost column, P denotes the class professor, and S denotes the class student. Pattern R2 is classified as P because out of the three nearest training patterns, majority belongs to the class professor: T2 and T4 are P, and T6 is S. For R3, the majority of the nearest training patterns belong to the class student, and hence R3 is classified as S.

reasoning. When you were asked in Section 1.3 to intuitively classify the professor–student recall patterns given in Fig. 1.2, it is quite likely that you instinctively carried out some case-based reasoning. On seeing a badly injured accident victim who needs immediate treatment, a doctor may examine the victim quickly and remembering a similar case successfully treated in the past, the doctor may prescribe the same class of treatment. Doctors gain experience by building a repertoire of past cases in their mind. For a given accident victim, different doctors with different experiences—different repertoires and different ways to assess similarity of cases—may prescribe differently.

6.4 Mixed Attribute Values

Often the attributes may be a mixture of numeric and non-numeric values. Given two patterns, we can calculate their weighted Euclidean distance [Eqn (6.3)] by using the numeric attributes, and the weighted Hamming distance by using the non-numeric attributes. We can then define the overall distance between the two patterns to be some function of their weighted Euclidean and Hamming distances. Instead of the weighted Euclidean distance, if we use the weighted city-block distance [Eqn (6.4)], then the overall distance between two patterns can be defined to be some function of their weighted city-block and Hamming distances.

Regardless of the attributes being numeric or non-numeric, classifying a recall pattern using the nearest neighbour classifier can require a lot of computation, since it requires calculating as many distances as there are training patterns. For most practical problems, the number of training patterns is expected to be large.

6.5 Prototype of a Class

To reduce computation while classification, each class in the training set can be represented by a prototype pattern. As the name suggests, the prototype is an estimate of the typical pattern in the class. For $1 \leq i \leq M$, if attribute A_i is numeric, then the value of A_i of the prototype can be taken to be the arithmetic mean of the A_i values of the training patterns of that class. Suppose a class has two training patterns: for the first pattern, $A_1 = 6$ and

$A_2 = 10$; for the second pattern, $A_1 = 4$ and $A_2 = 8$. Then, for the prototype

$$A_1 = \frac{6+4}{2} = 5$$

and

$$A_2 = \frac{10+8}{2} = 9$$

If attribute A_i is non-numeric, the value of A_i of the prototype can be taken to be the most common value of A_i in the training patterns of that class. This is a debatable way of estimating the attribute values of a prototype, but this is a design choice one may make depending on the domain of the patterns. After the prototypes have been obtained, a given recall pattern is then classified into the class of the nearest prototype, where the overall distance between the recall pattern and a prototype is some function of the Euclidean, city-block, and Hamming distances between the recall pattern and the prototype. This requires calculating as many distances as there are classes, which should be fewer than calculating as many distances as there are training patterns.

Consider the case where the attribute array $\bar{A} = A_1, A_2, \ldots, A_M$ has numeric values. The prototypes of the classes C_1 to C_m are represented by w_1 to w_m, where, for $1 \le k \le m$, the coordinates of prototype w_k are w_{k1}, $w_{k2}, \ldots, w_{kM}$ in the A_1-A_2- $\cdots$ -A_M coordinate space. Using Euclidean distance to classify a recall pattern with attribute array $\bar{A}$ to the class of the nearest prototype, we minimize

$$\sqrt{\sum_{i=1}^{M}(A_i - w_{ki})^2}$$

over $1 \le k \le m$. The square root being monotonic, we minimize

$$\sum_{i=1}^{M}(A_i - w_{ki})^2$$

That is, we minimize

$$\sum_{i=1}^{M}(A_i^2 - 2w_{ki}A_i + w_{ki}^2)$$

Since A_i^2 is a constant for all values of k, it can be removed, and hence we minimize

$$\sum_{i=1}^{M}(-2w_{ki}A_i + w_{ki}^2)$$

By changing signs, we instead maximize

$$\sum_{i=1}^{M}(2w_{ki}A_i - w_{ki}^2)$$

Dividing throughout by 2, we maximize

$$\sum_{i=1}^{M}(w_{ki}A_i - \frac{w_{ki}^2}{2})$$

Hence, we maximize

$$\sum_{i=1}^{M}w_{ki}A_i - \sum_{i=1}^{M}\frac{w_{ki}^2}{2} \qquad (6.5)$$

This is a linear equation in A_i. We refer to the equation as

$$g_k(\bar{A}) = w_{k0} + \sum_{i=1}^{M}w_{ki}A_i \qquad (6.6)$$

where

$$w_{k0} = -\sum_{i=1}^{M}\frac{w_{ki}^2}{2}$$

for $1 \leq k \leq m$. The coordinates of the prototype serve as weights w's, and $g_k(\bar{A})$ is a linear discriminant function for class C_k. To classify a pattern with its given attribute array $\bar{A}$, we evaluate the discriminant function for each class, and then assign the pattern to the class having the highest valued discriminant function. Each discriminant function is a hyperplane separating one class from another in the A_1-A_2- $\cdots$ -A_M coordinate space and the classes are linearly separable. As mentioned in Section 5.5, a classifier that uses linear discriminant functions is a linear classifier. Accordingly, a classifier that uses Euclidean distance to assign a recall pattern to the class of the nearest prototype can be formulated as a linear classifier (more on linear classifiers is given in Sections 7.2 and 7.3, and in Chapter 8).

As an example, suppose w_1, w_2, and w_3 are the prototypes of the three classes C_1, C_2, and C_3. The attribute array contains numeric attributes A_1 and A_2. So, $M = 2$. On the A_1-A_2 coordinate plane, the coordinates of w_1, w_2, and w_3 are, respectively, (2, 1), (5, 9), and (7, 3). That is, $w_{11} = 2$, $w_{12} = 1$, $w_{21} = 5$, $w_{22} = 9$, $w_{31} = 7$, and $w_{32} = 3$.

Let us classify a recall pattern with attributes $A_1 = 4$ and $A_2 = 6$. In other words, the coordinates of the recall pattern are (4, 6). We need to evaluate the discriminant function obtained from Eqns (6.5) and (6.6) for each of the three classes:

$$g_1(\bar{A}) = -\sum_{i=1}^{2} \frac{w_{1i}^2}{2} + \sum_{i=1}^{2} w_{1i} A_i = 11.5$$

$$g_2(\bar{A}) = -\sum_{i=1}^{2} \frac{w_{2i}^2}{2} + \sum_{i=1}^{2} w_{2i} A_i = 19.0$$

$$g_3(\bar{A}) = -\sum_{i=1}^{2} \frac{w_{3i}^2}{2} + \sum_{i=1}^{2} w_{3i} A_i = 17.0$$

Since the value of $g_2(\bar{A})$ is the largest, the recall pattern with $A_1 = 4$ and $A_2 = 6$ is classified into class C_2.

Let us now use Euclidean distance [Eqn (6.1)] for classifying the above recall pattern into the class of the nearest prototype to check whether the classification is identical. The Euclidean distance between the recall pattern with coordinates (4, 6) and prototype

$$w_1 \text{ is } \sqrt{(4-2)^2 + (6-1)^2} = \sqrt{29}$$

$$w_2 \text{ is } \sqrt{(4-5)^2 + (6-9)^2} = \sqrt{10}$$

$$w_3 \text{ is } \sqrt{(4-7)^2 + (6-3)^2} = \sqrt{18}$$

Since the recall pattern is nearest to the prototype w_2, it is classified into class C_2. So, the recall pattern's classification by the linear classifier and by minimizing the Euclidean distance is identical.

6.6   Performance of Nearest Neighbour Classifier

It is an established mathematical result that the error rate (proportion of patterns misclassified) of the nearest neighbour classifier will never be more than twice the error rate of the Bayes classifier (Chapter 5). The nearest neighbour error rate approaches the Bayes error rate in either of these situations:

(a) When the Bayes error rate approaches either 0 or 1
(b) When $k \to \infty$ and

$$\frac{\text{the number of training patterns}}{k} \to \infty$$

We can expect a nearest neighbour classifier to perform well when there are a lot of training patterns. However, the more training patterns there are, the more distances have to be calculated, and consequently the computation required increases, thus slowing down the process of classification.

6.7   Modifications in Nearest Neighbour Classifier

The following modifications can be adopted before implementing the nearest neighbour classifier:

1. If a training pattern T of class C has other training patterns of class C close to it, then remove T. A recall pattern that would have been classified by T will now be classified by the other training patterns that were close to T. So, the results are not expected to be affected. Similarly, if a training pattern T of class C has no other training patterns of class C close to it, then remove T, for T is perhaps an outlier that may cause a recall pattern to be misclassified. Repeatedly deleting training patterns like T is known as *editing* (or *condensing*, or *pruning*) the training set. Since an edited training set is smaller than the original training set, fewer distances need to be calculated when classifying a recall pattern, thus reducing the computation required, and consequently speeding up the process of classification.

2. In the k-nearest neighbour classifier, to the class of each of the k nearest neighbours, associate a weight that decreases as the distance d of the neighbour increases from the recall pattern. Sum up the weight

of each class among the k neighbours. Assign the recall pattern to the class with the greatest weight. If you have the weight as $1/d$, then the class of a neighbour with a small value of d tends to dominate the classes of the other neighbours. So a suggested weight is $1/(1 + d)$.

3. Find the $k \geq 1$ nearest neighbours from each class. Calculate the mean distance between the recall pattern and the k nearest neighbours of each class. Assign the recall pattern to the class having the least mean distance. The classifier is then known as $(k+k)$-nearest neighbour classifier.

Summary

Recall patterns are assigned the same class as their nearest training patterns. Euclidean or city-block distances can be calculated for classifying recall patterns having numeric attribute values; Hamming distance is calculated for recall patterns with non-numeric attribute values. The distance between two patterns depends on the values of the attributes of the patterns. A class may be represented by a prototype. When a recall pattern is classified into the class of the nearest prototype, the classifier can be formulated as a linear classifier.

Exercises

1. Write a computer program in a language of your choice to implement the k-nearest neighbour classification, where $k \geq 1$. The user should be able to define the value of k, or ask the program to suggest a value. In the latter case, the program should look through the training set, and then suggest a value of k (see Section 6.2 on how the program can suggest a value). The user should be able to accept that value, or override it with another value. The program should be able to work with both numeric and non-numeric attributes. The user should be able to define whether to use Euclidean, city-block, or Hamming distance; and what weights, if any, to put on the attributes. Moreover, the user should have the option to choose the method of classification (as described in Sections 6.2 and 6.3, or according to the second or third modifications mentioned in Section 6.7). Experiment with the

NAME of pattern	Values of attributes A_1	A_2	Class
Y1	3	15	C_1
Y2	6	15	C_3
Y3	12	10	C_2
Y4	9	17	C_3
Y5	1	6	C_1
Y6	14	4	C_2
Y7	10	11	C_2
Y8	11	8	C_2
Y9	2	4	C_1
Y10	5	16	C_3
Y11	4	10	C_1
Y12	13	6	C_2
Y13	5	7	C_1
Y14	7	14	C_3
Y1'	3	7	
Y2'	4	12	
Y3'	4	16	
Y4'	9	7	
Y5'	7	16	
Y6'	6	4	
Y7'	13	13	
Y8'	9	15	

FIGURE 6.2   Training and recall sets comprising patterns with two numeric attributes A_1 and A_2, and belonging to three classes C_1, C_2, and C_3. Patterns Y1 to Y14 constitute the training set, and patterns Y1' to Y8' constitute the recall set. The patterns can be plotted as points on the A_1-A_2 coordinate space to get an idea of how the patterns are scattered over the space.

various options in the program developed to do the following, and report your results.

(a) Classify the recall patterns given in Fig. 6.2. The training set is given in the same figure.

(b) Classify the recall patterns given in Fig. 6.3. The training set is given in the same figure.

(c) Classify the recall patterns of computer science departments described in the tenth exercise of Chapter 2.

(d) Classify the professor–student recall patterns given in Fig. 1.2. The training set is given in Fig. 1.1. Use weighted Hamming distance, putting the weight of attribute HABIT to be zero, and the weights of EATS and FOOTWEAR to be one each. Are the classifications same as those in Fig. 6.1, in which each of the attributes had a weight of one? If so, why is it so? (*Hint*: See Fig. 2.9)

2. Under what conditions would it be appropriate to have each class represented by a prototype (Section 6.5)? Base the answer on your intuition. Write a program in a language of your choice, which, given a training set, creates a prototype for each class. Later, it classifies a recall pattern into the class of its nearest prototype. Use the program to classify the recall patterns mentioned in parts (a), (b), and (c) of the first exercise. Compare your results with those obtained in the corresponding part in the first exercise.

NAME of pattern	Values of attributes		Class
	A_1	A_2	
Z1	1	4	C_1
Z2	2	−5	C_1
Z3	3	−2	C_2
Z4	4	−6	C_1
Z5	5	6	C_1
Z6	6	3	C_2
Z7	7	−7	C_1
Z8	8	−4	C_2
Z9	9	1	C_2
Z10	10	−3	C_2
Z11	11	−8	C_1
Z12	12	8	C_1
Z13	13	5	C_2
Z14	14	−9	C_1
Z15	15	9	C_1
Z16	16	2	C_2
Z1′	11	3	
Z2′	8	−6	
Z3′	7	−5	
Z4′	5	5	
Z5′	13	6	
Z6′	10	7	
Z7′	4	−5	
Z8′	6	4	

FIGURE 6.3 Training and recall sets comprising patterns with two numeric attributes A_1 and A_2, and belonging to two classes C_1 and C_2. Patterns Z1 to Z16 constitute the training set, and patterns Z1′ to Z8′ constitute the recall set. The patterns can be plotted as points on the A_1-A_2 coordinate space to get an idea of how the patterns are scattered over the space.

Learning Objectives

This chapter contains the following topics:

- a neural net comprising a network of neurodes
- a training procedure for classes that are linearly separable: examples on modelling AND and OR gates
- modelling an XOR gate
- a feedforward, strictly layered, completely connected architecture for a neural net
- activation functions: sigmoid and hyperbolic
- output functions: linear, threshold, and boundary
- forward propagation procedure
- training multilayer neural nets by backpropagation
- example on training and testing a neural net to recognize the professor–student patterns of the first chapter

7

Multilayer Neural Nets

7.1 Neurodes

Let us define a *neurode* (also called a 'perceptron' by some writers) to be a
processor—often, in practice, simulated by software—for which the output
is a previously defined function of the input. To illustrate, the input to the
neurode u of Fig. 7.1 comes from signal $x(0) = 1$ and from the real-valued
signals $x(1)$, $x(2), \ldots,$ $x(N)$, for some positive integer value of N. For $0 \leq$
$j \leq N$, each $x(j)$ flows through a *connector* to *reach* the neurode. Associated
with the connector through which $x(j)$ flows is a real number $w(j)$, which is
said to be the *weight* of the connector. The signal that the neurode receives is,
however, the signal that reaches the neurode multiplied by the weight of the
connector through which it reaches. The neurode thus receives the *weighted
signal* $w(j)x(j)$. Having $w(j) = 0$ is like having no connector through which
$x(j)$ can reach the neurode. Overall, the input I to the neurode is then the
sum of the weighted signals it receives, that is,

$$I = \sum_{j=0}^{N} w(j)x(j)$$
$$= w(0) + \sum_{j=1}^{N} w(j)x(j) \quad (\text{because} \;\; x(0) = 1)$$

As we shall soon see, having $x(0)$ is a design choice. If we have $x(0)$, then
fixing it to 1 is a convention. Even if all of $x(1)$ to $x(N)$ become zero, the
neurode still receives the non-zero input of $w(0)$, which is said to be the *bias*
received by the neurode. Having $w(0) = 0$ means having no $x(0)$. If there is
no $x(0)$, the neurode's input comes from $x(1)$ to $x(N)$.

The output y of the neurode can be defined to be some function of the input
I, for instance,

$$y = \begin{cases} 0 & \text{if } I < t \\ 1 & \text{otherwise} \end{cases}$$

where t is some threshold value. Later in this chapter, we will see how a
network of neurodes can be trained on the patterns such as the professor–
student patterns given in Fig. 1.1, and how the network can recognize the
corresponding recall patterns given in Fig. 1.2. Before that, however, we
need to be aware of a few simpler applications of neurodes, and why and
how neurodes are networked.

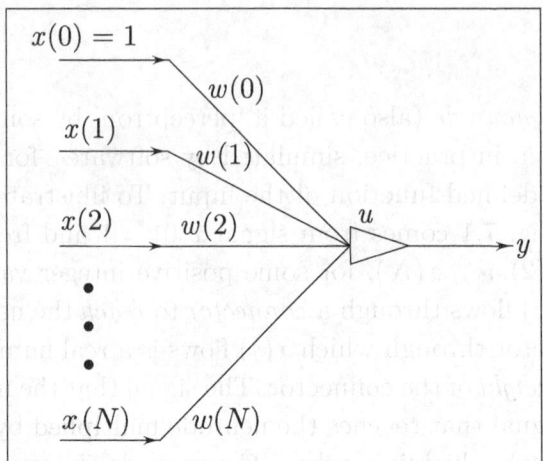

FIGURE 7.1   Real-valued signals $x(0)$ to $x(N)$ flow into neurode u, which outputs y. The weight of the connector through which $x(j)$ flows to u is $w(j)$, for $0 \leq j \leq N$. Overall, input I to u is equal to $\sum_{j=0}^{N} w(j)x(j)$. We can define a function so that output y is that function of input I.

7.2   Modelling an AND Gate

In its simplest form, a *neunet* (a contraction of *neural network*, a network of neurodes) consists of a single neurode such as in Fig. 7.1. Suppose we have one such single-neurode neunet, in which the input is coming from $x(1)$ and $x(2)$, the possible values for each of these being 0 and 1. With weights $w(1)$ and $w(2)$, input I to the neurode is $w(1)x(1) + w(2)x(2)$. The objective is to find the values of $w(1)$ and $w(2)$ so the neunet outputs 1 if both $x(1)$ and $x(2)$ are 1, and it outputs 0 otherwise: the neunet thus models an AND gate. The neunet is said to have *learned* the weights, or the neunet is said to have been *trained*.

Reverting to the terminology used in the earlier chapters, we can say that the neunet recognizes patterns from two classes: class YES-AND containing patterns whose attributes correspond to $x(1) = 1$ and $x(2) = 1$, and class NO-AND containing patterns whose attributes correspond to the remaining three permutations of the values of $x(1)$ and $x(2)$. The training set shown in Fig. 7.2, consists of the four patterns corresponding to the four possible permutations of the values of $x(1)$ and $x(2)$. There are no recall patterns distinct from the training patterns. The neunet is trained and tested on the same four patterns. Once the neunet has been trained, then by observing

Training pattern	Attributes corresponding to		Class	Desired output z of neunet
	$x(1)$	$x(2)$		
1	1	1	YES-AND	1
2	0	1	NO-AND	0
3	1	0	NO-AND	0
4	0	0	NO-AND	0

FIGURE 7.2   Training set for a neunet to model an AND gate. The desired output z of the neunet is different for the two classes, as it should be.

the output of the neurode for a given pattern, we can find out the class of the pattern. For instance, if the trained neurode outputs 1, then we can say that the pattern belongs to the class YES-AND.

An overview of the procedure to train a neunet is as follows. Begin with random weights of the connectors. Then one by one, process the training patterns: feed (the attribute values of) the training patterns to the neunet, and observe the neunet's output for each pattern. Note the error made by the neunet in recognizing the pattern. Update, or modify, the weights to reduce the error. The weights can be updated after processing every training pattern. Such updating of weights is said to be *case updating*, also known as 'online updating' or 'exemplar updating'. Alternatively, the weights can be updated after processing all the training patterns, such updating is known as *epoch updating* (an *epoch* is the processing of every training pattern once). In epoch updating, the updating of the weights is based on the accumulated error over all the training patterns. By either of the updating methods, the training may take many epochs: the training set may have to be processed many times, before the training can be considered to have been finished.

The procedure to train a neunet, such as the one shown in Fig. 7.1, is given below in more detail. Appropriate explanations have been given with the different steps illustrating how the steps are being adopted for training the neunet to model an AND gate. Understanding this procedure helps in easily understanding the more complicated training procedure given in Section 7.6 for a more elaborate, but practically more useful, neunet. The procedure to train a one-neurode neunet is the following.

1. The input to the neunet comes from real-valued signals $x(0)$, $x(1)$, ..., $x(N)$, for some positive integer value of N, where, if $x(0)$ is present, then, by convention, $x(0) = 1$. Decide whether to use $x(0)$ to provide a bias to the neurode, and on the value of N. To provide a bias by using $x(0)$, assign 0 to a variable B; otherwise assign 1 to B. The signals $x(1)$ to $x(N)$ represent the attribute values of the patterns on which the neunet is being trained. If the attribute values are non-numeric, they have to be coded so that the signals are numeric values. An example of such coding is given in Section 7.7. For modelling the AND gate, let us decide not to have any $x(0)$, and to have the value of N as 2. Accordingly, $B = 1$, and the input to the neunet comes from $x(1)$ and $x(2)$, each of which has possible values 0 and 1.

2. For each class, specify the output z the neunet should give. This is known as the *desired* output for the class. No two classes should have the same desired output. For the AND gate, let us specify that the desired output is $z = 1$ for patterns of class YES-AND and $z = 0$ for patterns of class NO-AND. These values of z are as shown in the rightmost column of the AND gate's training set given in Fig. 7.2.

3. Define output y of the neurode to be some function of the input I of the neurode. Then, y becomes the *observed* output of the neurode. The objective of training the neunet is to determine the weight $w(j)$ for $B \leq j \leq N$, so that, for the training patterns of each class, the observed output y matches the desired output z. Ideally, y *matches* z when $y = z$, but when y and z are real numbers, then, because of rounding off errors, y may never equal z, and hence, in practice, y is considered to match z when y is nearly equal to z within some margin of our choice. Once the training is finished, then on later observing some output y for a recall pattern, we can declare the pattern to belong to the class whose desired output was specified as the matching z. For modelling the AND gate, let us define

$$y = \begin{cases} 0 & \text{if } I < 0.8 \\ 1 & \text{otherwise} \end{cases}$$

4. Decide on whether case updating or epoch updating is to be used. For training the neunet to model an AND gate, let us decide to use case updating.

5. Fix a value of the *learning constant* $0 < \beta \leq 1$, to be used in step 20.1 to calculate the amount by which weights should be updated so

that the observed output y becomes closer to the desired output z. The value of β is a heuristic. A low value slows down the weight changes, thus slowing down the training, but a high value can result in overcorrection, thereby causing the weights to oscillate in successive changes. For noisy training patterns—those in which there are minor inaccuracies in recording the attribute values, for instance, 0.001 being recorded instead of 0—a low value of β is usually recommended. Let us choose $\beta = 0.2$ for training the neunet to model an AND gate.

6. Specify a value for epoch_bound, which indicates the expected number of epochs within which the training of the neunet will finish. The value of epoch_bound is a heuristic. For modelling the AND gate, let us choose the value of epoch_bound to be 10.

7. Let $|V|$ be the number of training patterns in training set V. For modelling the AND gate, as shown in the training set given in Fig. 7.2, the value of $|V|$ is 4.

8. Initialize $w(B)$ to $w(N)$ to non-zero, random real values, which will typically be a mixture of positive and negative values. Obtaining random values was discussed in Section 4.4. Only $w(1)$ and $w(2)$ need to be initialized for modelling the AND gate, for which, let us assume that $w(1)$ is initialized to 0.1, and $w(2)$ to 0.2.

9. epoch $\leftarrow 0$
The variable epoch is initialized to zero so it can be used to count the epochs, that is, to count the number of times each pattern of the training set is fed to the neunet.

10. epoch $\leftarrow$ epoch $+1$
Increment the counter for epochs by 1.

11. If epoch $>$ epoch_bound, then go to step 25 to report failure in training the neunet.

12. For $j = B$, 1, ..., N do
$s(j) \leftarrow w(j)$
Save the weights at the beginning of the epoch in array s. If the current epoch shows that the neunet is already trained, then, in step 22, the procedure will restore the weights to as they were at the beginning of the epoch.

13. If epoch updating was chosen in step 4, initialize an array $q(B)$ to $q(N)$ to zeroes. In steps 20.2 and 23, for $B \leq j \leq N$, element $q(j)$ will

serve as an accumulator to sum the amount by which $w(j)$ should be updated.

14. $T \leftarrow 0$

    The variable T is initialized to zero so that it can be used as a counter for the training patterns within each epoch.

15. $T \leftarrow T + 1$

    Increment the counter for the training patterns by 1.

16. If $T > |V|$, that is, all training patterns have been processed in the current epoch, then go to step 22 and check whether the neunet had been trained by the end of the last epoch.

17. Feed the $x(1)$ to $x(N)$ values of the Tth training pattern to the neunet.

18. $I \leftarrow \sum\limits_{j=B}^{N} w(j)x(j)$

    Thus, if $B = 0$ (there is bias), we sum from $j = 0$; otherwise, from $j = 1$. For modelling the AND gate, $B = 1$ as mentioned in step 1, hence input I is $w(1)x(1) + w(2)x(2)$.

19. Calculate the observed output y according to its definition in step 3.

20. For $j = B$, 1, ..., N do steps 20.1 and 20.2.

20.1. $\Delta w(j) \leftarrow \beta(z - y)x(j)$

      The amount by which $w(j)$ will be changed either in step 20.2 or in step 23 depends on $\Delta w(j)$ so that the error—the difference between the desired output z for the class of the Tth training pattern and the observed output y—is reduced, thereby bringing y closer to z. The error needs to be moderated by $x(j)$, the signal that flows through the connector of weight $w(j)$. It would then seem that $w(j)$ should be changed by $(z - y)x(j)$. In practice, this has been found to overcorrect $w(j)$: increased now, the weight may have to be decreased later, and then perhaps increased again. Thus, $w(j)$ may oscillate. To reduce the chance of this happening, $(z - y)x(j)$ is scaled down by multiplying it with the heuristic learning constant β, whose value, as selected in step 5, lies between 0 and 1. This formula for calculating $\Delta w(j)$ is known as the 'Widrow–Hoff rule' after B. Widrow and M.E. Hoff, who had proposed it.

20.2. If case updating is being used, then do

$$w(j) \leftarrow w(j) + \Delta w(j)$$

If, however, epoch updating is being used, then, for the current epoch, $q(j)$ accumulates the modifications required in $w(j)$, and accordingly, do

$$q(j) \leftarrow q(j) + \Delta w(j)$$

It has often been observed in practice that oscillations in weights are more likely with case updating than with epoch updating.

21. Go to step 15 to feed the neunet with the next training pattern in the current epoch.

22. For at least p per cent of the training patterns in the last epoch, if the difference between each observed output y and its corresponding desired output z was less than or equal to some heuristically chosen small value ϵ, then consider the training to be finished. We have to choose a value for p. For instance, if we choose p to be 95, then we are allowing 5% of the training patterns to be misclassified by the trained neunet. Ideally, the value for p should be 100, but sometimes we may have to accept a lower value for p because we may not be able to train the neunet to classify all training patterns correctly. If training has not finished, go to step 23; otherwise, the training having finished, the weights at the beginning of the epoch were the ones to be learned by the neunet. If epoch updating was chosen in step 4, then the weights have not changed in the current epoch, hence go to step 26 to return with the weights. If, however, we have done case updating in step 20.2, then the weights have changed during the epoch, and we have to restore the weights to as they were at the beginning of the epoch. We had saved the weights in array s in step 12. Therefore,

for $j = B, 1, \ldots, N$ do

$w(j) \leftarrow s(j)$

Go to step 26 to return with the weights.

23. If epoch updating was chosen in step 4, then

for $j = B, 1, \ldots, N$ do

$w(j) \leftarrow w(j) + \Delta q(j)$

The epoch has finished, and hence the weights are updated by the accumulated modification required for all of the patterns in the training set.

24. Go to step 10 to begin a new epoch.

25. Return with a message that the neunet failed to be trained within epoch_bound epochs. In such an event, one can try to train the neunet again by changing one or more of the following:
    - In step 1, if there was no bias provided to the neurode initially, then to provide it; or if there was a bias, then to remove it
    - The definition of the output function in step 3
    - The method of updating the weights chosen in step 4, case updating to epoch updating, or conversely
    - The value of the learning rate β selected in step 5
    - The value of the epoch_bound chosen in step 6
    - The initial random values of the weights $w(B)$ to $w(N)$ in step 8
    - The value of p in step 22
    - The value of ε in step 22

26. Return from the procedure with weights $w(B)$ to $w(N)$ that the neunet has learnt.

Using the above procedure, the training of a neunet to model an AND gate is shown in Fig. 7.3. The caption of the figure provides an explanation on how the training proceeds. The trained neunet is shown in Fig. 7.4, for which we have

$$w(1) = 0.5 \tag{7.1}$$

$$w(2) = 0.6 \tag{7.2}$$

$$I = w(1)x(1) + w(2)x(2) \tag{7.3}$$

$$y = 1 \text{ if } I \geq 0.8 \tag{7.4}$$

Substituting the values of $w(1)$, $w(2)$, and I in Eqns (7.3) and (7.4), we get output $y = 1$, if $0.5x(1) + 0.6x(2) \geq 0.8$. If we were to view each pattern in the training set given in Fig. 7.2 as a point on the $x(1)$-$x(2)$ coordinate plane (Fig. 7.5), then the straight line represented by the equation $0.5x(1) + 0.6x(2) = 0.8$ would divide the plane into two regions, corresponding to the two classes YES-AND and NO-AND: the two classes are linearly separable. The three points of the NO-AND class would be on one side of the line, and the single point of the YES-AND class would be on the other side. The line serves as a linear discriminant function to separate the patterns of the two

Epoch	$x(1)$	$x(2)$	Weights $w(1)$	$w(2)$	$I =$ $w(1)x(1)+$ $w(2)x(2)$	y	z	$\Delta w(j) =$ $\beta(z-y)x(j)$ $\Delta w(1)$	$\Delta w(2)$
1	1	1	0.1	0.2	0.3	0	1	0.2	0.2
	0	1	0.3	0.4	0.4	0	0	0.0	0.0
	1	0	0.3	0.4	0.3	0	0	0.0	0.0
	0	0	0.3	0.4	0.0	0	0	0.0	0.0
2	1	1	0.3	0.4	0.7	0	1	0.2	0.2
	0	1	0.5	0.6	0.6	0	0	0.0	0.0
	1	0	0.5	0.6	0.5	0	0	0.0	0.0
	0	0	0.5	0.6	0.0	0	0	0.0	0.0
3	1	1	0.5	0.6	1.1	1	1	0.0	0.0
	0	1	0.5	0.6	0.6	0	0	0.0	0.0
	1	0	0.5	0.6	0.5	0	0	0.0	0.0
	0	0	0.5	0.6	0.0	0	0	0.0	0.0

FIGURE 7.3 Training of a neunet to model an AND gate. As shown in the first row of epoch 1, we have $w(1) = 0.1$ and $w(2) = 0.2$ by random initialization. With $x(1) = 1$ and $x(2) = 1$, input $I = w(1)x(1) + w(2)x(2) = (0.1 \times 1) + (0.2 \times 1) = 0.3$. According to the function defined for output in step 4 of the training procedure, since $I < 0.8$, observed output $y = 0$. The desired output z is, however, 1. Defining learning constant β to be equal to 0.2, we get $\Delta w(1) = \beta(z - y)x(1) = 0.2 \times (1 - 0) \times 1 = 0.2$. Similarly, $\Delta w(2) = \beta(z - y)x(2) = 0.2 \times (1 - 0) \times 1 = 0.2$. Using case updating, each of $w(1)$ and $w(2)$ is increased by 0.2. Thus, $w(1)$ becomes $0.1 + 0.2 = 0.3$, and $w(2)$ becomes $0.2 + 0.2 = 0.4$. This is shown in the second row of epoch 1 in the columns for $w(1)$ and $w(2)$. The training of the neunet proceeds similarly row by row. The training is finished after the third epoch when no changes occur in the weights $w(1)$ and $w(2)$, the difference between the desired output and the observed output having become 0 for each training pattern. The training finishes with $w(1) = 0.5$ and $w(2) = 0.6$. The trained neunet is shown in Fig. 7.4.

classes, and the neunet works as a linear classifier (more on linear classifiers in Chapter 8).

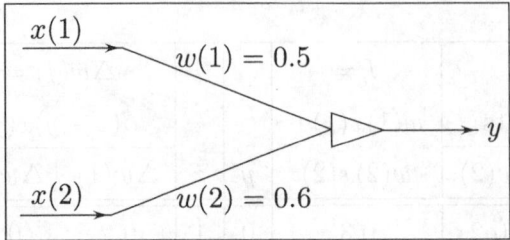

FIGURE 7.4 A neunet modelling an AND gate. Input I to the neurode is $w(1)x(1) + w(2)x(2)$, where each of $x(1)$ and $x(2)$ can be either 0 or 1. By definition, output y is 0 if $I < 0.8$; otherwise, y is 1. With the weights as shown, y becomes 1 if both $x(1)$ and $x(2)$ are 1; otherwise, y becomes 0. You may want to confirm this by reading through the last four rows of the table in Fig. 7.3.

As should be apparent from Fig. 7.5, there can be more than one straight line that can divide the coordinate plane in this manner. For instance, the line represented by $0.3x(1) + 0.7x(2) = 0.8$ will also divide the plane into the same two regions. The neunet could have finished its training with $w(1) = 0.3$ and $w(2) = 0.7$. The values of $w(1)$ and $w(2)$ for which the training finishes can depend on the initial random values of the weights and on the value chosen for the learning constant β.

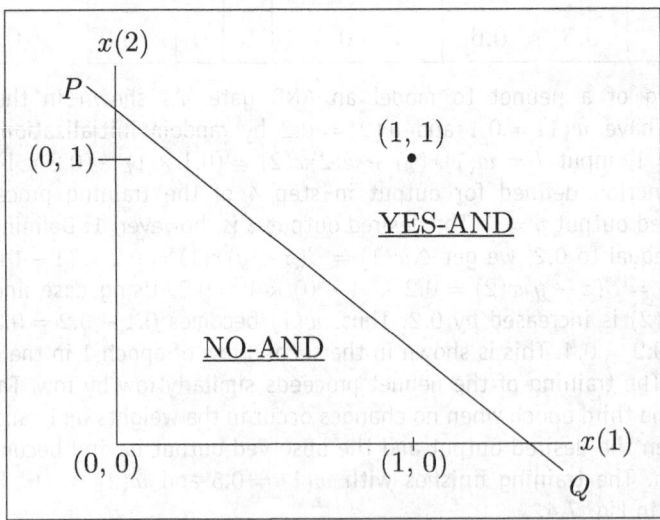

FIGURE 7.5 Representing the AND gate model on the $x(1)$-$x(2)$ coordinate plane. A straight line PQ divides the four points into two regions corresponding to the two classes YES-AND and NO-AND.

We can find out the constraints on the weights in terms of the threshold t used in the output function, according to which

$$y = \begin{cases} 0 & \text{if } I < t \\ 1 & \text{otherwise} \end{cases}$$

For the first pattern in the training set given in Fig. 7.2, output y should be 1; for that to happen, input I should be greater than or equal to t, that is,

$$[w(1)x(1) + w(2)x(2)] \geq t \tag{7.5}$$

since $I = w(1)x(1) + w(2)x(2)$. From the second to the fourth patterns of Fig. 7.2, output y should be 0, for which, input I should be less than t, that is,

$$[w(1)x(1) + w(2)x(2)] < t \tag{7.6}$$

Substituting the values of $x(1)$ and $x(2)$ in Eqns (7.5) and (7.6), we obtain the following four equations:

$$[w(1) \times 1 + w(2) \times 1] \geq t \text{ (from the first pattern)} \tag{7.7}$$

$$[w(1) \times 0 + w(2) \times 1] < t \text{ (from the second pattern)} \tag{7.8}$$

$$[w(1) \times 1 + w(2) \times 0] < t \text{ (from the third pattern)} \tag{7.9}$$

$$[w(1) \times 0 + w(2) \times 0] < t \text{ (from the fourth pattern)} \tag{7.10}$$

Simplifying Eqns (7.7) to (7.9), we get the following constraints on the weights $w(1)$ and $w(2)$ for a neunet to model an AND gate

$$[w(1) + w(2)] \geq t \quad [\text{from Eqn (7.7)}]$$
$$w(2) < t \quad [\text{from Eqn (7.8)}]$$
$$w(1) < t \quad [\text{from Eqn (7.9)}]$$

Figure 7.3 shows that, with $t = 0.8$, the neunet finished its training at $w(1) = 0.5$ and $w(2) = 0.6$. Clearly, the weights satisfy the above constraints. As mentioned earlier, with different initial weights and a different value of the learning constant β, the neunet could have finished its training at, say, $w(1) = 0.3$ and $w(2) = 0.7$. These weights, too, would satisfy the above constraints. In general, the weight values at which a neunet finishes training need not be unique.

7.3   Modelling an OR Gate

Just as a neunet consisting of a single neurode, as shown in Fig. 7.4, can be trained to model an AND gate, similarly it can also be trained to model an OR gate. The input comes from $x(1)$ and $x(2)$, the possible values for each of these being 0 and 1. The neunet outputs 0 if both $x(1)$ and $x(2)$ are 0; otherwise it outputs 1. Employing terminology similar to that in Section 7.2, class NO-OR contains patterns whose attributes correspond to $x(1) = 0$ and $x(2) = 0$, and class YES-OR contains patterns whose attributes correspond to the remaining three permutations of the values of $x(1)$ and $x(2)$. The training set for a neunet to model an OR gate is shown in Fig. 7.6.

The neunet can be trained by applying the training procedure described in Section 7.2. Since the training of the AND gate is shown in some detail in Fig. 7.3, the training of the OR gate has not been similarly shown. Let the output y of the neurode be

$$y = \begin{cases} 0 & \text{if } I < 0.5 \\ 1 & \text{otherwise} \end{cases}$$

The neunet can be shown to finish its training (Fig. 7.7) having

$$w(1) = 0.7 \tag{7.11}$$

$$w(2) = 0.5 \tag{7.12}$$

Training pattern	Attributes corresponding to		Class	Desired output z of neunet
	$x(1)$	$x(2)$		
1	1	1	YES-OR	1
2	0	1	YES-OR	1
3	1	0	YES-OR	1
4	0	0	NO-OR	0

FIGURE 7.6   Training set for a neunet to model an OR gate.

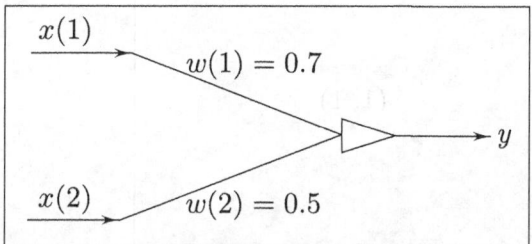

FIGURE 7.7 A neunet modelling an OR gate. Input I to the neunet is $w(1)x(1) + w(2)x(2)$, where each of $x(1)$ and $x(2)$ can be either 0 or 1. By definition, output y of the neurode is 0 if $I < 0.5$; otherwise, y is 1. With the weights as shown, y becomes 0 if both $x(1)$ and $x(2)$ are 0; otherwise, y becomes 1.

$$I = w(1)x(1) + w(2)x(2) \tag{7.13}$$

$$y = 1 \text{ if } I \geq 0.5 \tag{7.14}$$

Substituting the values of $w(1)$, $w(2)$, and I in Eqns (7.13) and (7.14) we get output $y = 1$ if $0.7x(1) + 0.5x(2) \geq 0.5$. On the $x(1)$-$x(2)$ coordinate plane (Fig. 7.8), the straight line represented by the equation $0.7x(1) + 0.5x(2) = 0.5$ divides the plane into two regions, corresponding to the two classes YES-OR and NO-OR: the two classes are linearly separable. The line serves as a linear discriminant function to separate the patterns of the two classes, and the neunet works as a linear classifier.

As should be apparent from Fig. 7.8, there can be more than one straight line that can divide the coordinate plane in this manner. For instance, the line represented by $0.6x(1) + 0.6x(2) = 0.5$ will also divide the plane into the same two regions. Therefore, as observed in Section 7.2, the weights at which the AND gate finishes its training are not unique; similarly the weights at which the OR gate finishes its training are not unique either. The neunet for modelling the OR gate could have finished its training at $w(1) = 0.6$ and $w(2) = 0.6$.

In Section 7.2, we had obtained the constraints on weights for a neunet modelling an AND gate. Similarly, for an OR gate, we can obtain the constraints on the weights in terms of the threshold t used in the output function, according to which

$$y = \begin{cases} 0 \text{ if } I < t \\ 1 \text{ otherwise} \end{cases}$$

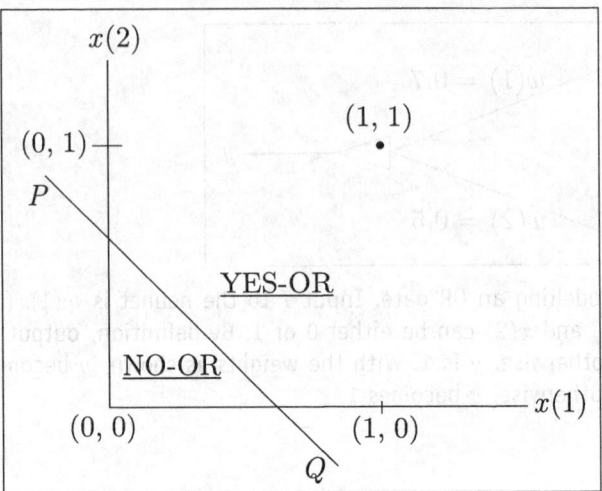

FIGURE 7.8 Representing the OR gate model on the $x(1)$-$x(2)$ coordinate plane. A straight line PQ divides the four points into two regions corresponding to the two classes YES-OR and NO-OR.

For the first three patterns of the training set given in Fig. 7.6, output y should be 1; hence input I should be greater than or equal to t, that is,

$$[w(1)x(1) + w(2)x(2)] \geq t \tag{7.15}$$

since $I = w(1)x(1) + w(2)x(2)$. For the fourth pattern of Fig. 7.6, output y should be 0; hence input I should be less than t, that is,

$$[w(1)x(1) + w(2)x(2)] < t \tag{7.16}$$

Substituting the values of $x(1)$ and $x(2)$, in Eqns (7.15) and (7.16), we obtain the following four equations:

$$[w(1) \times 1 + w(2) \times 1] \geq t \quad \text{(from the first pattern)} \tag{7.17}$$

$$[w(1) \times 0 + w(2) \times 1] \geq t \quad \text{(from the second pattern)} \tag{7.18}$$

$$[w(1) \times 1 + w(2) \times 0] \geq t \quad \text{(from the third pattern)} \tag{7.19}$$

$$[w(1) \times 0 + w(2) \times 0] < t \quad \text{(from the fourth pattern)} \tag{7.20}$$

Simplifying Eqns (7.17) to (7.19), we get the following constraints on the weights $w(1)$ and $w(2)$ for a neunet to model an OR gate:

$$[w(1) + w(2)] \geq t \quad \text{[from Eqn (7.17)]}$$

$$w(2) \geq t \quad \text{[from Eqn (7.18)]}$$
$$w(1) \geq t \quad \text{[from Eqn (7.19)]}$$

Figure 7.7 shows that, with $t = 0.5$, the neunet finished its training at $w(1) = 0.7$ and $w(2) = 0.5$. Clearly, the weights satisfied the above constraints. The neunet could have also finished its training at, say, both $w(1)$ and $w(2)$ being 0.6.

7.4   Modelling an XOR Gate

In Sections 7.2 and 7.3, we saw that single-neurode neunets can be trained to model AND and OR gates. Suppose we have to try training such a neunet to model an XOR (Exclusive OR) gate. The input to the neunet comes from $x(0,1)$ and $x(0,2)$, the possible values for each of these being 0 and 1—having 0 as the first argument of x may seem unnecessary, but it is being used to conform to notation needed when we will build the XOR-gate neunet, and for neunets in general as described in Section 7.5. The neunet will output 0 if both $x(0,1)$ and $x(0,2)$ are identical, and it will output 1 otherwise. In terminology compatible with that in Sections 7.2 and 7.3, class NO-XOR will contain patterns whose attributes corresponding to $x(0,1)$ and $x(0,2)$ are identical, and class YES-XOR otherwise. The training set for a neunet to model an XOR gate is in Fig. 7.9.

Training pattern	Attributes corresponding to		Class	Desired output z of neunet
	$x(0,1)$	$x(0,2)$		
1	1	1	NO-XOR	0
2	0	1	YES-XOR	1
3	1	0	YES-XOR	1
4	0	0	NO-XOR	0

FIGURE 7.9   Training set used for a neunet to model an XOR gate.

Suppose, for some value of t, we define output y of the neurode as

$$y = \begin{cases} 0 & \text{if } I < t \\ 1 & \text{otherwise} \end{cases}$$

For the first and fourth patterns of the training set given in Fig. 7.9, output y should be 0, hence input I should be less than t, that is,

$$[w(1)x(0,1) + w(2)x(0,2)] < t \tag{7.21}$$

since $I = w(1)x(0,1) + w(2)x(0,2)$. For the second and third patterns of Fig. 7.9, output y should be 1; therefore input I should be greater than or equal to t, that is,

$$[w(1)x(0,1) + w(2)x(0,2)] \geq t \tag{7.22}$$

Substituting the values of $x(0,1)$ and $x(0,2)$ in Eqns (7.21) and (7.22), we obtain the following four equations:

$$[w(1) \times 1 + w(2) \times 1] < t \text{ (from the first pattern)} \tag{7.23}$$

$$[w(1) \times 0 + w(2) \times 1] \geq t \text{ (from the second pattern)} \tag{7.24}$$

$$[w(1) \times 1 + w(2) \times 0] \geq t \text{ (from the third pattern)} \tag{7.25}$$

$$[w(1) \times 0 + w(2) \times 0] < t \text{ (from the fourth pattern)} \tag{7.26}$$

Simplifying Eqns (7.23) to (7.25), we get the following constraints on the weights $w(1)$ and $w(2)$ for a neunet to model an XOR gate

$$[w(1) + w(2)] < t \quad \text{[from Eqn (7.23)]}$$

$$w(2) \geq t \quad \text{[from Eqn (7.24)]}$$

$$w(1) \geq t \quad \text{[from Eqn (7.25)]}$$

Adding Eqns (7.24) and (7.25), we get

$$[w(1) + w(2)] \geq 2t$$

But this contradicts Eqn (7.23), which says $[w(1) + w(2)] < t$. Therefore, an XOR gate cannot be modelled by a one-neurode neunet.

Figure 7.10 illustrates that, on the $x(0,1)$-$x(0,2)$ coordinate plane, there does not exist a straight line that would divide the plane into two regions

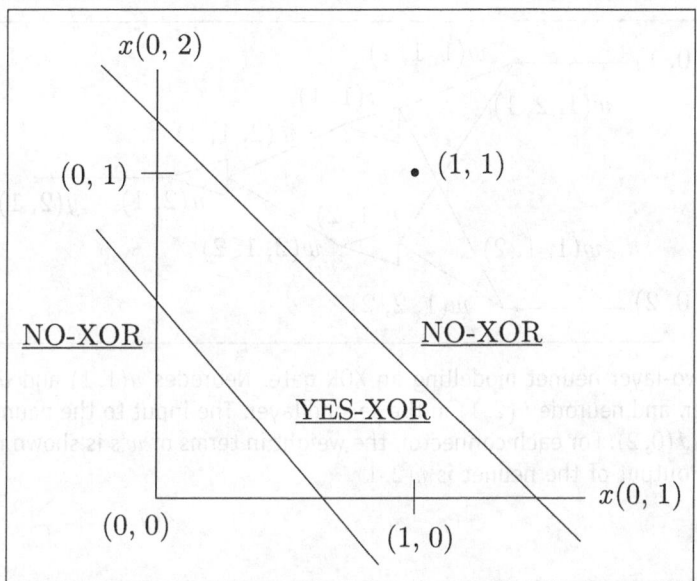

FIGURE 7.10 On the $x(0,1)$-$x(0,2)$ coordinate plane, there does not exist a straight line to separate the YES-XOR class from the NO-XOR class. Points (0, 1) and (1, 0) belong to the YES-XOR class, and points (0, 0) and (1, 1) belong to the NO-XOR class.

corresponding to the two classes YES-XOR and NO-XOR: the two classes are not linearly separable. It is an established result that classes which are not linearly separable cannot be recognized by a neunet containing only one neurode.

For an XOR gate, we need a non-linear discriminant function. The XOR gate can be modelled by a neunet that contains three neurodes as shown in Fig. 7.11. Its output for the four permutations of input values is given in Fig. 7.12, whose caption explains, as an example, how the values in the first row of the table are calculated (readers may want to refer frequently to the two figures while reading the description below). The notation required here for the XOR gate is an extension of the notation for the AND and OR gates of Sections 7.2 and 7.3. Understanding the description in this section by the example of an XOR gate should help in understanding the more general neunet described in Section 7.5.

The neunet of Fig. 7.11 has two *layers* of neurodes: neurodes $u(1,1)$ and $u(1,2)$ in the first layer, and $u(2,1)$ in the second layer. The first argument of u indicates the layer of the neurode and the second argument indicates the sequential count of the neurode in that layer. As mentioned above, the input

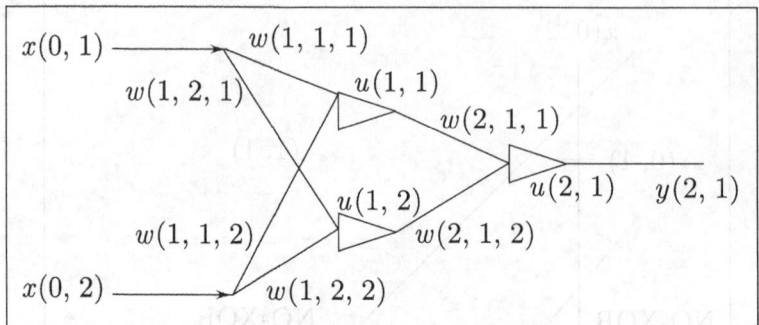

FIGURE 7.11 A two-layer neunet modelling an XOR gate. Neurodes $u(1,1)$ and $u(1,2)$ are in the first layer, and neurode $u(2,1)$ in the second layer. The input to the neunet comes from $x(0,1)$ and $x(0,2)$. For each connector, the weight in terms of w's is shown next to the connector. The output of the neunet is $y(2,1)$.

Training Pattern	$x(0,1)$ that is $x(1,1)$	$x(0,2)$ that is $x(1,2)$	$I(1,1)$	$I(1,2)$	$y(1,1)$ that is $x(2,1)$	$y(1,2)$ that is $x(2,2)$	$I(2,1)$	$y(2,1)$
1	1	1	0	0	0	0	0	0
2	0	1	-1	1	0	1	1	1
3	1	0	1	-1	1	0	1	1
4	0	0	0	0	0	0	0	0

FIGURE 7.12 Outputs of the neunet shown in Fig. 7.11 modelling an XOR gate, for the four possible permutations of $x(0,1)$ and $x(0,2)$. Before entering the first-layer neurodes, signal $x(0,1)$ is renamed $x(1,1)$, and $x(0,2)$ is renamed $x(1,2)$. In the first row, for instance, given $x(1,1)=1$ and $x(1,2)=1$, we employ Eqn (7.27) by which $I(1,1)=w(1,1,1)\times x(1,1)+w(1,1,2)\times x(1,2)$, where $w(1,1,1)=1$ and $w(1,1,2)=-1$, to get $I(1,1)=0$. We again employ Eqn (7.27) by which $I(1,2)=w(1,2,1)\times x(1,1)+w(1,2,2)\times x(1,2)$, where $w(1,2,1)=-1$ and $w(1,2,2)=1$, to get $I(1,2)=0$. According to Eqn (7.28), since $I(1,1)<0.1$, output from $u(1,1)$ is $y(1,1)=0$; according to the same equation, since $I(1,2)<0.1$, output from $u(1,2)$ is $y(1,2)=0$. Outputs $y(1,1)$ and $y(1,2)$ are, respectively, renamed as inputs $x(2,1)$ and $x(2,2)$ to neurode $u(2,1)$. According to Eqn (7.29), we have $I(2,1)=w(2,1,1)\times x(2,1)+w(2,1,2)\times x(2,2)$, where $w(2,1,1)=1$ and $w(2,1,2)=1$; hence we get $I(2,1)=0$. According to Eqn (7.30), since $I(2,1)<0.1$, output from $u(2,1)$ is $y(2,1)=0$, which is the desired output from the neunet, as given in the first row of Fig. 7.9. Similar calculations can be made for the other rows in this figure.

to the neunet comes from $x(0,1)$ and $x(0,2)$, the possible values of each of which are 0 and 1. The points at which signals $x(0,1)$ and $x(0,2)$ split (the arrow tips on the two incoming x's in Fig. 7.11) are considered here to be the zeroth layer of neurodes $u(0,1)$ and $u(0,2)$, although no neurodes are required there (they are not shown in Fig. 7.11) since no computation is done at these points. Making this assumption, however, simplifies the notation: the first argument of signal x indicates the layer of the neurode into which x is flowing; moreover, $w(L,k,j)$ can then denote the weight of the connector from neurode $u(L-1,j)$ to $u(L,k)$, for $L \geq 1$. The weight of each connector in Fig. 7.11 is shown in terms of the w's written next to the connector.

Convention differs in the way the layers in a neunet are counted: by not counting the zeroth layer, some writers would call the neunet given in Fig. 7.11 to be a two-layer neunet; others would count the zeroth layer and call the neunet to be a three-layer neunet. The former practice is preferred in this book—the neurodes in the zeroth layer are notional; they are imagined to be present only for notational simplicity. Accordingly, each of the neunets shown in Figs 7.1, 7.4, and 7.7 has one layer and the neunet given in Fig. 7.11 has two layers.

A neunet with a single layer can have more than one neurode in that layer, the output of no neurode flowing into another neurode in the same layer. It has been mentioned earlier in this section that it is an established result that classes which are not linearly separable cannot be recognized by a neunet containing only one neurode. In fact, the result is more general: classes which are not linearly separable cannot be recognized by a neunet containing only one layer of neurodes. For such classes—they are the ones often found in practice—we need neunets of more than one layer, as we do for the XOR-gate neunet shown in Fig. 7.11, in which signal $x(0,1)$ flows into $u(0,1)$, and $x(1,2)$ into $u(0,2)$. Since no computation is done at these two u's, the output of each u is equal to its input: thus, the outputs from $u(0,1)$ and $u(0,2)$ are, respectively, $x(0,1)$ and $x(0,2)$.

Both $x(0,1)$ and $x(0,2)$ then split and—without any change in their values—reach neurode $u(1,1)$, and similarly reach neurode $u(1,2)$. To conform to our notation that the first argument of x indicates the layer of the neurode into which x is flowing, after the x's split, we rename $x(0,1)$ to be $x(1,1)$ and $x(0,2)$ to be $x(1,2)$. Inputs $I(1,1)$ and $I(1,2)$ into neurodes $u(1,1)$ and

$u(1, 2)$, respectively, are then obtained from the matrix equation:

$$\begin{bmatrix} I(1,1) \\ I(1,2) \end{bmatrix} = \begin{bmatrix} w(1,1,1) = 1 & w(1,1,2) = -1 \\ w(1,2,1) = -1 & w(1,2,2) = 1 \end{bmatrix} \begin{bmatrix} x(1,1) \\ x(1,2) \end{bmatrix} \tag{7.27}$$

In this matrix equation, an entry such as $w(1,1,1) = 1$ denotes that in the neunet shown in Fig. 7.11, weight $w(1,1,1)$ has value 1. The procedure that can be used to train the neunet so that $w(1,1,1)$ takes value 1 is described in Section 7.5. For now, it should be sufficient to see that the neunet does model an XOR gate, as explained in the caption of Fig. 7.12. For $1 \leq k \leq 2$, the output from neurode $u(1, k)$ is defined to be

$$y(1, k) = \begin{cases} 0 & \text{if } I(1, k) < 0.1 \\ 1 & \text{otherwise} \end{cases} \tag{7.28}$$

The outputs from the first layer of neurodes flow into the neurode in the second layer. Thus, the input to neurode $u(2, 1)$ comes from $y(1, 1)$ and $y(1, 2)$. To maintain uniformity in notation that input to a neurode is denoted by x's whose first argument indicates the layer of the neurode into which x is flowing, we rename $y(1, 1)$ to be $x(2, 1)$, and $y(1, 2)$ to be $x(2, 2)$. Input $I(2, 1)$ to neurode $u(2, 1)$ is then obtained from the matrix equation:

$$\begin{bmatrix} I(2,1) \end{bmatrix} = \begin{bmatrix} w(2,1,1) = 1 & w(2,1,2) = 1 \end{bmatrix} \begin{bmatrix} x(2,1) \\ x(2,2) \end{bmatrix} \tag{7.29}$$

The output from neurode $u(2, 1)$ is defined to be

$$y(2, 1) = \begin{cases} 0 & \text{if } I(2, 1) < 0.1 \\ 1 & \text{otherwise} \end{cases} \tag{7.30}$$

which is the output of the neunet. Figure 7.12 shows that the output $y(2, 1)$ of the neunet shown in Fig. 7.11 is always identical to the corresponding desired output as given in Fig. 7.9. Accordingly, we can say that the neunet of Fig. 7.11 models an XOR gate.

Figure 7.13 shows an alternative neunet which also models an XOR gate. The neurodes in Fig. 7.11 do not receive any bias (Section 7.1 explained what a bias is), whereas neurodes $u(1, 1)$, $u(1, 2)$, and $u(2, 1)$ in Fig. 7.13 receive biases. The notation for the neunet shown in Fig. 7.13 is the same as that for the neunet shown in Fig. 7.11.

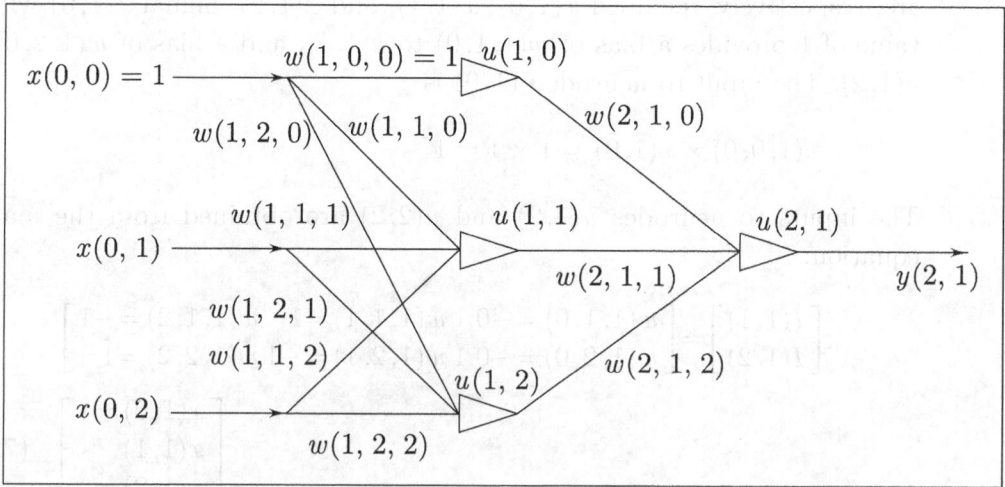

FIGURE 7.13 A neunet modelling an XOR gate in which the neurodes receive biases. Compare this neunet with the one shown in Fig. 7.11, in which the neurodes did not receive any bias.

Training Pattern	$x(0,1)$ that is $x(1,1)$	$x(0,2)$ that is $x(1,2)$	$I(1,1)$	$I(1,2)$	$y(1,1)$ that is $x(2,1)$	$y(1,2)$ that is $x(2,2)$	$I(2,1)$	$y(2,1)$
1	1	1	-0.1	-0.1	0	0	-0.1	0
2	0	1	-1.1	0.9	0	1	0.9	1
3	1	0	0.9	-1.1	1	0	0.9	1
4	0	0	-0.1	-0.1	0	0	-0.1	0

FIGURE 7.14 Output of the neunet shown in Fig. 7.13 for the four possible permutations of $x(0,1)$ and $x(0,2)$. The calculations for each of the rows are similar to those given in the caption of Fig. 7.12. The values of $I(1,1)$ and $I(1,2)$ are obtained from Eqn (7.31), the values of $y(1,1)$ and $y(1,2)$ from Eqn (7.32), the value of $I(2,1)$ from Eqn (7.33), and the value of $y(2,1)$ from Eqn (7.34).

The input to the neunet shown in Fig. 7.13 comes from $x(0,0)$, $x(0,1)$, and $x(0,2)$, out of which $x(0,0)$ is fixed at 1. At the zeroth layer, the incoming signals split. Their values, however, remain unchanged. To conform to our notation for the x's, after the x's split, signals $x(0,0)$, $x(0,1)$, and $x(0,2)$, are, respectively, renamed $x(1,0)$, $x(1,1)$, and $x(1,2)$. Signal $x(1,0)$ with a value of 1 provides a bias of $w(1,1,0)$ to $u(1,1)$, and a bias of $w(1,2,0)$ to $u(1,2)$. The input to neurode $u(1,0)$ is

$$w(1,0,0) \times x(1,0) = 1 \times 1 = 1$$

The inputs to neurodes $u(1,2)$ and $u(2,2)$ are obtained from the matrix equation:

$$\begin{bmatrix} I(1,1) \\ I(1,2) \end{bmatrix} = \begin{bmatrix} w(1,1,0)=-0.1 \ w(1,1,1)=1 \ \ w(1,1,2)=-1 \\ w(1,2,0)=-0.1 \ w(1,2,1)=-1 \ w(1,2,2)=1 \end{bmatrix}$$
$$\begin{bmatrix} x(1,0)=1 \\ x(1,1) \\ x(1,2) \end{bmatrix} \qquad (7.31)$$

No computation being done at biasing neurode $u(1,0)$, it outputs $y(1,0)=1$ to provide a bias of $w(2,1,0)$ to neurode $u(2,1)$. For $1 \le k \le 2$, the output from neurode $u(1,k)$ is defined to be

$$y(1,k) = \begin{cases} 0 \text{ if } I(1,k) < 0 \\ 1 \text{ otherwise} \end{cases} \qquad (7.32)$$

Output $y(1,0)$, together with outputs $y(1,1)$ and $y(1,2)$, flows into neurode $u(2,1)$. For uniformity in notation as explained above, $y(1,0)$ is renamed $x(2,0)$. Similarly, $y(1,1)$ and $y(1,2)$ are, respectively, named $x(2,1)$ and $x(2,2)$. Then the input to neurode $u(2,1)$ is

$$\big[I(2,1) \big] = \big[w(2,1,0)=-0.1 \ w(2,1,1)=1 \ w(2,1,2)=1 \big]$$
$$\begin{bmatrix} x(2,0)=1 \\ x(2,1) \\ x(2,2) \end{bmatrix} \qquad (7.33)$$

The output from neurode $u(2,1)$ is defined to be

$$y(2,1) = \begin{cases} 0 \text{ if } I(2,1) < 0 \\ 1 \text{ otherwise} \end{cases} \qquad (7.34)$$

which is considered to be the output of the neunet. Figure 7.14 shows that the output $y(2,1)$ of the neunet shown in Fig. 7.13 is always identical to the corresponding desired output given in Fig. 7.9. Accordingly, it can be said that the neunet shown in Fig. 7.13 models an XOR gate just as the neunet shown in Fig. 7.11 does.

7.5 Commonly used Neunet Architecture

There are several *architectures* (how neurodes are connected to one another) in neunets. The most commonly used architecture, shown in Fig. 7.15, is a generalization of the XOR-gate neunet architectures given in Figs 7.11 and 7.13, with details as described below. Since a foreglimpse of the architecture and the notation is covered in Section 7.4 about XOR-gate neunets, the readers should find the description below easy to understand. Although the notation here is similar to that for XOR-gate neunets, it is more elaborate. While reading the description below, readers can frequently refer to Fig. 7.15 to get an idea of the layout of the neunet being discussed. A summary description of notations is also given in Fig. 7.16.

The neunet contains $G \geq 1$ layers, which are numbered $0, 1, 2, \ldots, G$. Every layer $0 \leq L \leq G$ contains neurodes $u(L, 1), u(L, 2), \ldots, u[L, N(L)]$, where $N(L)$ is an element of array N, such that each element of the array has a value of 1 or more. Layers $L = 0, 1, \ldots, (G-1)$ may, in addition, each contain a neurode $u(L, 0)$ called the *biasing neurode*. The layer G does not contain any biasing neurode. Neurodes that are not biasing are called *non-biasing neurodes*. Since $N(L) \geq 1$, each layer has at least one non-biasing neurode. It will be assumed for this description that biasing neurodes are present. In case of neunets that do not have any biasing neurodes, ignore whatever is said about biasing neurodes. In layer $0 \leq L < G$, if a variable ranges over 0 to $N(L)$ with biasing neurodes, then it ranges over 1 to $N(L)$ without biasing neurodes. As we did in Section 7.2, let us have variable $B = 0$ if biasing neurodes are present, and $B = 1$ otherwise. The neunet shown in Fig. 7.15 depicts a *feedforward*, *strictly layered*, and *completely connected* neunet.

Feedforward: Signals flow from neurodes in layer L to neurodes in layers greater than L, but not to neurodes in L itself, or to neurodes in layers less than L, for $0 \leq L < G$. Signal from a neurode in layer G does not flow to any other neurode. It is considered to be an output of the neunet.

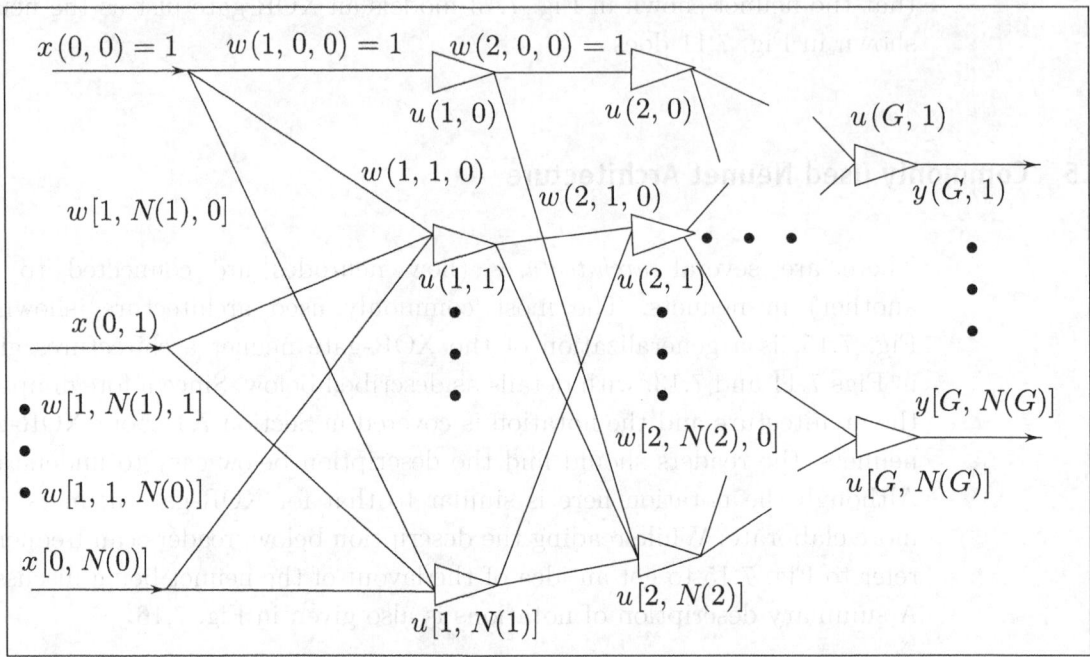

FIGURE 7.15 The most commonly used architecture of a neunet. As explained in Section 7.5, the neunet is feedforward, strictly layered, and fully connected. The u's are neurodes, the x's are signals flowing through the connectors, the w's are the weights of the connectors, and the y's are the outputs.

Strictly layered Signals flow from layer L only to layer $(L + 1)$, for $0 \le L < G$. Thus, for instance, although the neunet is feedforward, signals do not flow from layer L to, say, layer $(L + 2)$.

Completely connected For $0 \le L < G$, (a) every non-biasing neurode in layer L is connected to send signals to every non-biasing neurode in layer $(L + 1)$, a non-biasing neurode in layer L does not send a signal to the biasing neurode in layer $(L + 1)$; and (b) the biasing neurode in layer L is connected to send signals to every neurode in layer $(L + 1)$, the biasing neurode in layer L thus providing a bias to the neurodes in layer $(L + 1)$.

The objective of this chapter is to use neunets for recognizing patterns of the kind discussed in the earlier chapters. The neurodes in the zeroth layer (also known as the *input layer*) receive signals $x(0, 0), x(0, 1), \ldots, x[0, N(0)]$, of which $x(0, 0)$ with a fixed value of 1 flows into the biasing neurode and the

G	output layer in the neunet with layers numbered $0, 1, \ldots, G$.
B	indicates whether the neunet has biasing neurodes ($B = 0$) or does not have biasing neurodes ($B = 1$).
$N(L)$	number of non-biasing neurodes in layer $0 \leq L \leq G$.
$u(L, j)$	neurode j in layer $0 \leq L < G$, for $B \leq j \leq N(L)$. When $j = 0$, then $u(L, j)$ is a biasing neurode, otherwise a non-biasing neurode. The output layer has no biasing neurode; hence when $L = G$, we have $1 \leq j \leq N(G)$.
$w(L, k, j)$	weight of the connector from neurode $u(L - 1, j)$ to neurode $u(L, k)$, for layer $1 \leq L \leq G$, where $1 \leq k \leq N(L)$ and $B \leq j \leq N(L-1)$. Furthermore, $w(L, 0, 0) = 1$ is the weight of the connector from biasing neurode $u(L-1, 0)$ to biasing neurode $u(L, 0)$, for layer $1 \leq L < G$.
$x(L, j)$	signal that flows from neurode $u(L-1, j)$ to the neurodes of layer $1 \leq L \leq G$, where $B \leq j \leq N(L-1)$, with the constraint that no signal flows from a non-biasing neurode in layer $L-1$ to a biasing neurode in layer L, since no connector exists between these neurodes. Moreover, $x(0, 0)$, $x(0, 1), \ldots, x[0, N(0)]$ flow into the zeroth layer. By convention, $x(L, 0) = x(L-1, 0) = \cdots = x(0, 0) = 1$ since these signals provide the bias.
$I(L, k)$	input to neurode $u(L, k)$, for $1 \leq L \leq G$ and $B \leq k \leq N(L)$.
$a(L, j)$	activation value of neurode $u(L, j)$, for $0 \leq L \leq G$, where $B \leq j \leq N(L)$.
$y(G, k)$	observed output of neurode $u(G, k)$, for $1 \leq k \leq N(G)$.
$z(G, k)$	desired output of neurode $u(G, k)$, for $1 \leq k \leq N(G)$.

FIGURE 7.16 A summary of the notation described in Section 7.5 for a neunet such as the one in Fig. 7.15.

remaining signals, representing the attribute values of some given pattern, flow into the non-biasing neurodes.

Thereafter, the signal coming out from every neurode $u(L, k)$ in layer $0 \leq L < G$ is split to flow through weighted connectors to the neurodes of layer

$(L + 1)$. The signal is not weakened by splitting: what reaches a neurode in layer $(L + 1)$ is equal to the output of $u(L, k)$. The layer G is the *output layer*; its output is the output of the neunet, and this output indicates the class of the pattern whose attribute values were input to the zeroth layer. Layers $1, 2, \ldots, (G - 1)$ are neither the input, nor the output layers. These are known as the *hidden layers*. In an abstract sense, the hidden layers try to partition the space of different permutations of attribute values into subspaces, one subspace for each class of patterns.

As discussed in Section 7.4, although the zeroth layer does not need any neurodes since no computation is done at that layer, by assuming that neurodes are present in the zeroth layer, we simplify the notation: $x(L, j)$ then denotes the signal that flows from neurode $u(L - 1, j)$ to the neurodes of layer $1 \leq L \leq G$, for $B \leq j \leq N(L - 1)$, with the constraint that no signal flows from a non-biasing neurode in layer $L - 1$ to a biasing neurode in layer L, since no connector exists between these neurodes.

Signals $x(0, 0) = 1$, $x(0, 1)$, $\ldots$, $x[0, N(0)]$ flowing into the neurodes of the zeroth layer flow on unchanged to the neurodes of the first layer, since the output of every zero-layer neurode is the same as its input, no computation is done at the neurode. To conform to the notation that the first argument of x indicates the layer of the neurode into which x is flowing, for $1 \leq j \leq N(0)$, signal $x(0, j)$ is renamed $x(1, j)$ to denote it flowing into the neurodes of the first layer.

The weight of the connector from neurode $u(L - 1, j)$ to neurode $u(L, k)$, for layer $1 \leq L \leq G$, where $1 \leq k \leq N(L)$, and $B \leq j \leq N(L - 1)$, is $w(L, k, j)$. Since there is no connector from a non-biasing neurode in layer $L - 1$ to a biasing neurode in layer L, the range of k begins from 1, not 0, in the previous sentence. Furthermore, $w(L, 0, 0) = 1$ is designated to be the weight of the connector from biasing neurode $u(L - 1, 0)$ to biasing neurode $u(L, 0)$, for layer $1 \leq L < G$. Having no connector from a non-biasing neurode in layer $L - 1$ to a biasing neurode in layer L could have been alternatively expressed by saying $w(L, 0, j) = 0$ for $1 \leq L < G$ and $1 \leq j \leq N(L - 1)$. If we had not assumed neurodes in layer 0, we would have to use a different notation for connectors going from layer 0 to the non-biasing neurodes of layer 1, thus complicating the overall notation.

The input to neurode $u(L, k)$ is denoted by $I(L, k)$, for $1 \leq L \leq G$ and $B \leq k \leq N(L)$. As mentioned in Section 7.1, the input to a neurode is equal to the sum of the weighted signals that it receives. Accordingly, the input to a biasing neurode $u(L, 0)$ is

$$I(L, 0) = w(L, 0, 0)\ x(L, 0) = 1 \times 1 = 1$$

for $1 \leq L < G$. No computation is done at a biasing neurode. The output of a biasing neurode is equal to its input. As a result, signal $x(L, 0)$, flowing from biasing neurode $u(L - 1, 0)$ to biasing neurode $u(L, 0)$, is always equal to 1.

For a non-biasing neurode $u(L, k)$, that is, for $1 \leq L \leq G$ and $1 \leq k \leq N(L)$, input

$$I(L, k) = \sum_{j=B}^{N(L-1)} w(L, k, j)\ x(L, j) \tag{7.35}$$

The equation becomes clearer if we look at its matrix form in Fig. 7.17. As the figure's caption explains, when $B = 0$, the neunet has to learn $N(L) \times [N(L - 1) + 1]$ weights between the layers $(L - 1)$ and L. The neunet has to learn weights between layers 0 and 1, between layers 1 and 2, and so on, till between layers $(G - 1)$ and G. Therefore, the total number of weights the neunet has to learn is

$$\sum_{L=1}^{G} N(L) \times [N(L - 1) + 1] \tag{7.36}$$

As an example, consider the XOR-gate neunet of Fig. 7.13, which has $G = 2$, $N(0) = 2$, $N(1) = 2$, and $N(2) = 1$ (count only the non-biasing neurodes). The total number of weights the neunet needs to learn is

$$\sum_{L=1}^{2} N(L) \times [N(L - 1) + 1]$$
$$= [N(1) \times (N(0) + 1)] + [N(2) \times (N(1) + 1)]$$
$$= (2 \times 3) + (1 \times 3)$$
$$= 9$$

It can be confirmed from Fig. 7.13 that the neunet would need to learn the nine weights $w(1, 1, 0)$, $w(1, 1, 1)$, $w(1, 1, 2)$, $w(1, 2, 0)$, $w(1, 2, 1)$, $w(1, 2, 2)$, $w(2, 1, 0)$, $w(2, 1, 1)$, and $w(2, 1, 2)$. The weight $w(1, 0, 0)$ need not be

$$
\begin{bmatrix}
I(L,1) \\
I(L,2) \\
I(L,3) \\
\vdots \\
I[L,N(L)]
\end{bmatrix}
=
\begin{bmatrix}
w(L,1,0) & w(L,1,1) & \cdots & w[L,1,N(L-1)] \\
w(L,2,0) & w(L,2,1) & \cdots & w[L,2,N(L-1)] \\
w(L,3,0) & w(L,3,1) & \cdots & w[L,3,N(L-1)] \\
\vdots & \vdots & \ddots & \vdots \\
w[L,N(L),0] & w[L,N(L),1] & \cdots & w[L,N(L),N(L-1)]
\end{bmatrix}
$$

$$
\times
\begin{bmatrix}
x(L,0)=1 \\
x(L,1) \\
x(L,2) \\
\vdots \\
x[L,N(L-1)]
\end{bmatrix}
$$

FIGURE 7.17 Matrix equation to evaluate $I(L,k)$, the input to neurode $u(L,k)$, for $1 \leq L \leq G$ and $1 \leq k \leq N(L)$, when biasing neurodes are present. The equation can be reformulated in a non-matrix form as $I(L,k) = \sum_{j=0}^{N(L-1)} w(L,k,j)\ x(L,j)$. The matrix equation shows that between the layers $(L-1)$ and L, the neunet has to learn the weight matrix of $N(L)$ rows and $[N(L-1)+1]$ columns. If biasing neurodes are not present, then delete the first column of the weight matrix since each of the weights $w(L,1,0)$, $w(L,2,0)$, ..., $w[L,N(L),0]$ will be zero. Moreover, delete $x(L,0)$ from the x array since there will be no signal being input to the biasing neurode. Hence, without biasing neurodes, the neunet will need to learn a weight matrix of $N(L)$ rows and $N(L-1)$ columns. In a non-matrix form, the equation can then be written as $I(L,k) = \sum_{j=1}^{N(L-1)} w(L,k,j)\ x(L,j)$. The above two non-matrix-form equations for $I(L,k)$ can be merged into one equation as $I(L,k) = \sum_{j=B}^{N(L-1)} w(L,k,j)\ x(L,j)$, which is shown as Eqn (7.35) in Section 7.5.

counted, since the neunet does not have to learn it: as discussed above and as the figure shows, its value is fixed at 1.

The caption of Fig. 7.17 clarifies that, when $B = 1$, the neunet has to learn $N(L) \times N(L-1)$ weights between the layers $(L-1)$ and L. The total number of weights the neunet has to learn is then

$$
\sum_{L=1}^{G} N(L) \times N(L-1) \tag{7.37}
$$

Consider, as an example, the XOR-gate neunet shown in Fig. 7.11, which has $G = 2$, $N(0) = 2$, $N(1) = 2$, and $N(2) = 1$. The total number of weights the neunet needs to learn is

$$\sum_{L=1}^{2} N(L) \times N(L-1)$$
$$= [N(1) \times N(0)] + [N(2) \times N(1)]$$
$$= (2 \times 2) + (1 \times 2)$$
$$= 6$$

It can be confirmed from Fig. 7.11 that the neunet would need to learn the six weights $w(1,1,1)$, $w(1,1,2)$, $w(1,2,1)$, $w(1,2,2)$, $w(2,1,1)$, and $w(2,1,2)$.

Subtracting Eqn (7.37) from Eqn (7.36), we notice that the neunet has to learn

$$\sum_{L=1}^{G} N(L) \tag{7.38}$$

more weights when biasing neurodes are present than when they are not.

A neurode is said to reach an *activation value* when it receives signals. The activation values of neurodes of layer L flow into the neurodes of layer $(L+1)$, for $1 \leq L < G$. The activation value of a zero-layer neurode and of a biasing neurode is equal to the input of that neurode.

For a non-biasing neurode in layers 1 to G, we need to define a function $\mathcal{A}$ so we can calculate the neurode's activation value from its input I. The function can be discrete; for instance, the activation value can depend on whether I is above some threshold value. This, however, results in some loss of accuracy. Accordingly, it is usually preferred to have continuous activation values, since they provide greater accuracy. It is an established mathematical result that, to train multilayer neunets by the backpropagation procedure (described in Section 7.6), the activation function must be monotonically increasing and differentiable. The following two activation functions, which furnish continuous values, are often used.

- *Sigmoid, or logistic, function*

$$\mathcal{A}(I) = \frac{1}{1 + e^{-cI}} \tag{7.39}$$

Often c is chosen to be 1. At $I = 0$, we get $\mathcal{A}(I) = 0.5$. Moreover,

$$\lim_{I \to -\infty} \mathcal{A}(I) \to 0$$

$$\lim_{I \to \infty} \mathcal{A}(I) \to 1$$

Thus, $\mathcal{A}(I)$ lies between 0 and 1, regardless of the value of I. This prevents the activation value from ever becoming so large as to overflow the capacity of computation of the neunet. The derivative $\mathcal{A}'$ of $\mathcal{A}(I)$ with respect to I is

$$\mathcal{A}'(I) = \frac{ce^{-cI}}{(1 + e^{-cI})^2}$$

$$= \frac{c(1 + e^{-cI} - 1)}{(1 + e^{-cI})^2}$$

$$= \frac{c}{1 + e^{-cI}} \left(1 - \frac{1}{1 + e^{-cI}} \right)$$

$$= c\mathcal{A}(I)[1 - \mathcal{A}(I)] \text{ [from Eqn (7.39)]}$$

For $c = 1$, the plots of $\mathcal{A}(I)$ and its derivative are shown in Fig. 7.18.

- *Hyperbolic function* $\tanh(I)$

$$\mathcal{A}(I) = \tanh(I)$$

At $I = 0$, we get $\mathcal{A}(I) = 0$. Moreover,

$$\lim_{I \to -\infty} \tanh(I) \to -1$$

$$\lim_{I \to \infty} \tanh(I) \to 1$$

Thus, no matter what the value of I is, $\tanh(I)$ lies between -1 and 1. As it is for the sigmoid function, this prevents the activation value from ever becoming so large as to overflow the capacity of computation of the neunet. The plots of $\tanh(I)$ and its derivative $\text{sech}^2(I)$ are shown in Fig. 7.19.

Let $a(L, j)$ be the activation value of neurode $u(L, j)$. Summarizing this discussion, it can be said that the activation value

- $a(0, j)$ of neurode $u(0, j)$, for $0 \leq j \leq N(0)$ (that is, for a neurode in the zeroth layer), is equal to $x(0, j)$, the input to the neurode; a special case of this is $a(0, 0) = 1$, because $x(0, 0) = 1$ by convention;

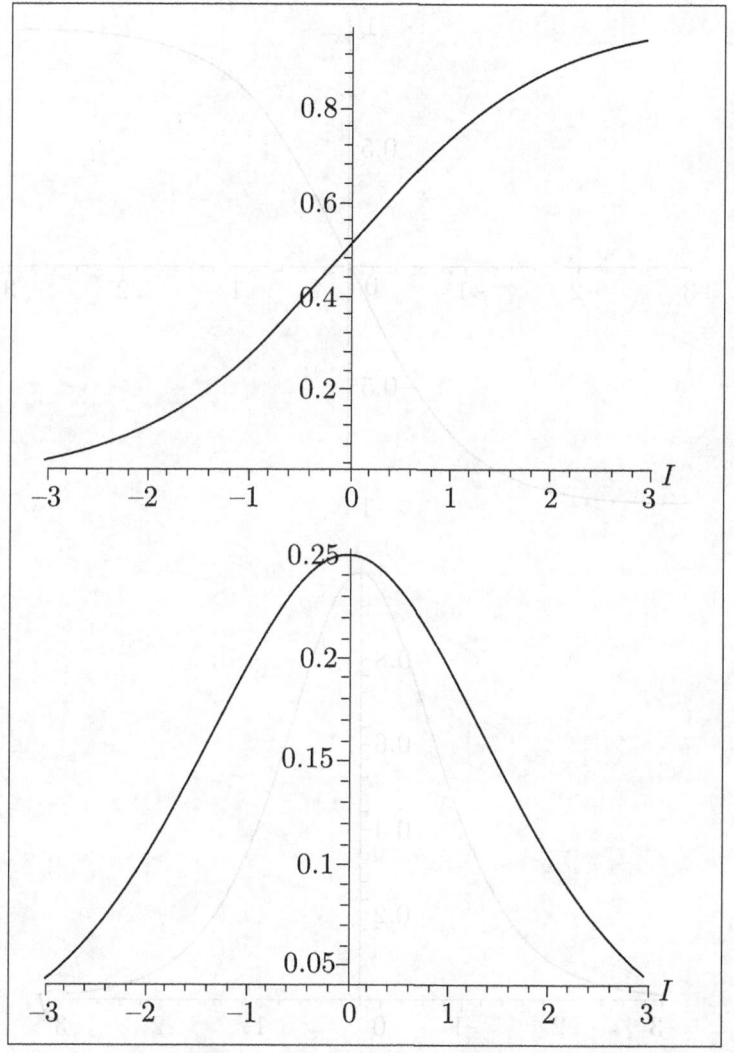

FIGURE 7.18 Curves obtained by plotting the sigmoid function $1/(1 + e^{-I})$ at the top and its derivative $e^{-I}/(1 + e^{-I})^2$ below.

- $a(L, 0)$ of neurode $u(L, 0)$, for $0 \leq L < G$ (that is, for a biasing neurode in layers 0 to $G - 1$), is 1 by convention;
- $a(L, k)$ of neurode $u(L, k)$, for $1 \leq L \leq G$, and $1 \leq k \leq N(L)$ (that is, for a non-biasing neurode in layers 1 to G), is obtained by applying the activation function $\mathcal{A}$ to the input $I(L, k)$ of $u(L, k)$; for instance, if the

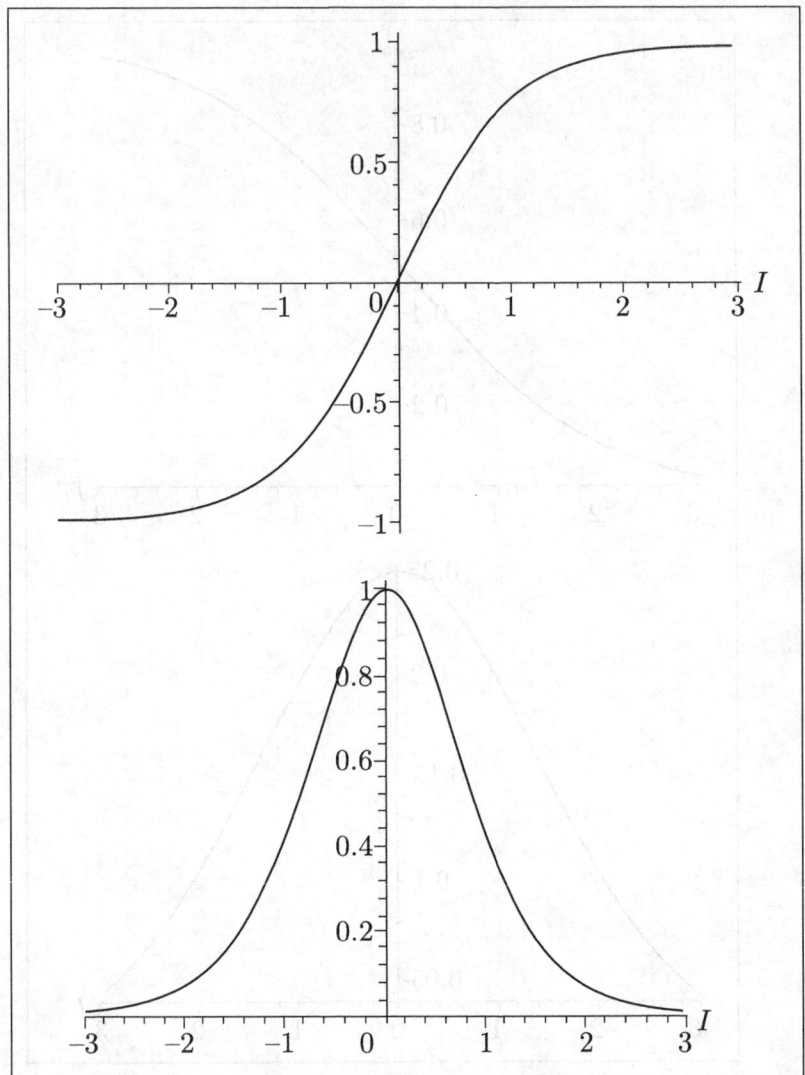

FIGURE 7.19 Curves obtained by plotting the hyperbolic function $\tanh(I)$ at the top, and its derivative $\text{sech}^2(I)$ below.

activation function is chosen to be sigmoid with its constant $c = 1$, then

$$a(L, k) = \frac{1}{1 + e^{-I(L,k)}}$$

Although, we can have a different activation function for every non-biasing neurode, it is customary to have the same activation function for all such neurodes in a given neunet.

We had earlier designated $x(L, j)$ to be the signal that flows from neurode $u(L-1, j)$ to the neurodes of layer L. Because it is the activation value of $u(L-1, j)$ that flows into the neurodes of layer L, we can say that

$$x(L, j) = a(L-1, j)$$

for $1 \leq L \leq G$ and $B \leq j \leq N(L-1)$.

Even with continuous activation values, we may find it convenient to have a discrete output from layer G, so it can be displayed for us to see. For $1 \leq k \leq N(G)$, the discrete output $y(G, k)$ of a neurode $u(G, k)$ is a function of the activation value $a(G, k)$ of the neurode. Since y is being defined only for layer G, G could have been omitted from the arguments of y. However, G is being put as an argument of y, just to maintain consistency with the notation for u, w, x, I, a, whose first argument specifies a layer.

It is an established mathematical result that to train multilayer neunets by the backpropagation procedure of Section 7.6, just as an activation function must be monotonically increasing and differentiable, so must an output function be. The three commonly used output functions, for $1 \leq k \leq N(G)$, are the following:

- *Linear function*

$$y(G, k) = b_1 a(G, k) + b_2$$

  where b_1 and b_2 are constants we choose. If $b_1 = 1$ and $b_2 = 0$, then the function becomes the identity function $y(G, k) = a(G, k)$.

- *Threshold function*

$$y(G, k) = \begin{cases} 0 & \text{if } a(G, k) < t \\ v & \text{otherwise} \end{cases}$$

  where t is a threshold value, and v is some value. So $y(G, k)$ has a value of 0 or v. We choose the values for t and v, although typically $t < 1$ and $v = 1$.

- *Boundary function*

$$y(G, k) = \begin{cases} 0 & \text{if } a(G, k) < t_1 \\ a(G, k) & \text{if } t_1 \leq a(G, k) \leq t_2 \\ 1 & \text{otherwise} \end{cases}$$

where t_1 and t_2 are threshold values we choose. Output value $y(G, k)$ is thus in the closed interval [0,1]. Suppose we use the sigmoid activation function, then $a(G, k)$ would never become 0 or 1. With the boundary output function, when $a(G, k) < t_1$, that is, $a(G, k)$ is close to 0, the neurode would output 0. Similarly, when $a(G, k) > t_2$, that is, $a(G, k)$ is close to 1, the neurode would output 1. Typically, $t_1 = 0.1$ and $t_2 = 0.9$.

In an informal sense, think of activation as the anger one feels inside when annoyed by something or someone. Output is how one expresses it: appear calm, yell, swear, or hit out.

Suppose we have a neunet for which the following values are known: G, $N(0)$, $N(1)$, ..., $N(G)$; the weights of the different connectors; the activation function $\mathcal{A}$; the output function $\mathcal{O}$ at layer G; whether the neunet has biasing neurodes $(B = 0)$, or not $(B = 1)$; and the values of the signals $x(0, 1)$, $x(0, 2)$, ..., $x[0, N(0)]$ flowing into the zeroth layer.

The following *forward propagation procedure* shows how to trace the signals flowing from layer to layer so that the value of the input $I(L, k)$ into every non-biasing neurode can be calculated; and the values of the outputs $y(G, 1)$, $y(G, 2)$, ..., $y[G, N(G)]$ of layer G can be calculated.

1. If $B = 0$, then do
   for $L = 0, 1, \ldots, G$
   $x(L, 0) \leftarrow 1$
   The signal flowing into every biasing neurode is 1.

2. For $j = 1, 2, \ldots, N(0)$ do
   $a(0, j) \leftarrow x(0, j)$
   The signal flowing into a neurode in the zeroth layer becomes the activation value of the neurode.

3. For $L = 1, 2, \ldots, G$ do steps 3.1 and 3.2.

   3.1. For $j = 1, 2, \ldots, N(L - 1)$ do
   $x(L, j) \leftarrow a(L - 1, j)$
   The activation values of the neurodes in layer $(L - 1)$ become the signals flowing into layer L.

   3.2. For $k = 1, 2, \ldots, N(L)$, do steps 3.2.1 and 3.2.2.

   3.2.1. $I(L, k) \leftarrow \sum_{j=B}^{N(L-1)} w(L, k, j) \, x(L, j)$

$I(L, k)$, the input to neurode $u(L, k)$, is obtained by using Eqn (7.35)

3.2.2. Apply activation function $\mathcal{A}$ to $I(L, k)$ to obtain $a(L, k)$, the activation value of neurode $u(L, k)$:

$$a(L, k) \leftarrow \mathcal{A}[I(L, k)]$$

4. For $k = 1, 2, \ldots, N(G)$ do

$$y(G, k) \leftarrow \mathcal{O}[a(G, k)]$$

Output function $\mathcal{O}$ is applied to $a(G, k)$, the activation value of neurode $u(G, k)$ in layer G, to obtain the neurode's output $y(G, k)$.

5. Return with the values of $I(L, k)$, for $1 \leq L \leq G$ and $1 \leq k \leq N(L)$. Also, return with the values of $y(G, 1), y(G, 2), \ldots, y[G, N(G)]$. All these values will be needed by the backpropagation procedure to train multilayer neunets, described in Section 7.6.

7.6 Training Multilayer Neunets by Backpropagation

All the notations presented in the last section must have been overwhelming. Figure 7.16 is useful to know what is denoted by G, B, $N(L)$, $u(L, k)$, $w(L, k, j)$, $x(L, j)$ $I(L, k)$, $a(L, j)$, $y(G, k)$, and $z(G, k)$, used in this section.

The backpropagation procedure to train a multilayer neunet is an extension of the procedure to train a one-neurode neunet described in Section 7.2. We feed training patterns one by one to the neunet, and then we update the weights depending on the error. The backpropagation procedure is said to perform a 'gradient descent' on the error surface in the weight space: the weights change in the direction of steepest descent on the error surface. In other words, the weights change in the direction of the greatest rate of decrease of error.

In a multilayer neunet, for a given training pattern, error $e(G, k)$ in the output of neurode $u(G, k)$, for $1 \leq k \leq N(G)$, is defined to be

$$e(G, k) = \mathcal{A}'[I(G, k)] \times [z(G, k) - y(G, k)] \tag{7.40}$$

in which $[z(G, k) - y(G, k)]$ is the difference between the desired output $z(G, k)$ and the observed output $y(G, k)$. The derivative $\mathcal{A}'[I(G, k)]$ gives the rate of change of the activation function $\mathcal{A}$ at $I(G, k)$, the value input to neurode $u(G, k)$. The derivative scales the difference $[z(G, k) - y(G, k)]$.

The slower the activation function changes at $I(G, k)$, the lower the value of $\mathcal{A}'[I(G, k)]$, and consequently the lower is the value of $e(G, k)$.

Since the layers $0 \leq L < G$ are not output layers, the desired output of neurodes in these layers is not known. Error $e(L, j)$ in the output of neurode $u(L, j)$, where $B \leq j \leq N(L)$, in these layers is sent to all neurodes in layer $(L + 1)$. Thus, $e(L, j)$ contributes to error $e(L + 1, k)$ of neurode $u(L + 1, k)$, for $1 \leq k \leq N(L + 1)$, that is, $e(L, j)$ contributes to the error of each of the neurodes in layer $(L + 1)$. Error $e(L, j)$ is distributed over the neurodes of layer $(L + 1)$ in proportion to the weight of the connector between $u(L, j)$ and the different neurodes. The derivative $\mathcal{A}'[I(L, j)]$—which gives the rate of change of the activation function $\mathcal{A}$ at $I(L, j)$, the value input to $u(L, j)$—is used to scale the error; hence $e(L, j)$ is defined to be

$$e(L, j) = \mathcal{A}'[I(L, j)] \sum_{k=1}^{N(L+1)} w(L + 1, k, j)\, e(L + 1, k) \qquad (7.41)$$

for $0 \leq L < G$ and $B \leq j \leq N(L)$. Thus, errors in the neurodes of layer L can be obtained from the errors in the neurodes of layer $(L + 1)$.

To train a multilayer neunet, we commence by evaluating errors in the output layer G, knowing which we can evaluate the errors in layer $(G - 1)$, which in turn we use to evaluate the errors in layer $(G - 2)$, and so on, till we evaluate the errors in layer 1. As an outcome of this, we iteratively update weights starting from output layer G, then layer $(G - 1)$, then $(G - 2)$, and so on, till layer 1; that is, we first update all $w(G, k, j)$'s, then all $w(G - 1, k, j)$'s, and so on, till we update all $w(1, k, j)$'s. In other words, we first update the weights of the connectors between layers $(G - 1)$ and G, then between layers $(G - 2)$ and $(G - 1)$, and so on, till we update the weights of the connectors between layers 0 and 1.

In step 20.1 of the training procedure for a one-neurode neunet (Section 7.2), the amount of change required in weight $w(j)$ depended on $\Delta w(j) = \beta(z - y)x(j)$, where β was the learning constant with a user-selected value between 0 and 1, the difference between the desired output z and the observed output y was the error, and $x(j)$ was the signal flowing through the connector whose weight was $w(j)$. Correspondingly, for training a multilayer neunet, we ought to have $\Delta w(L, k, j) = \beta e(L, k)x(L, j)$, for $1 \leq L \leq G$, where $1 \leq k \leq N(L)$ and $B \leq j \leq N(L - 1)$. This is, however, sometimes not enough.

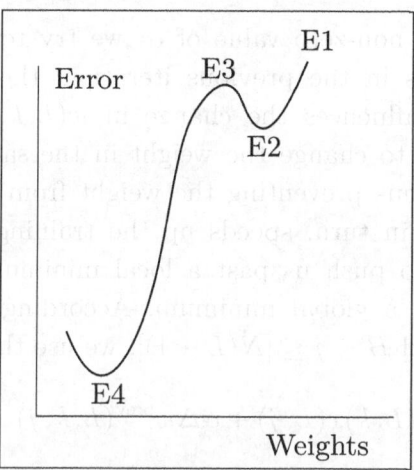

FIGURE 7.20 A plot of error on the weight plane. Point E2 is a local minimum and E4 is the global minimum.

Suppose there are D weights to be updated in the neunet. These can be viewed as a D-dimensioned weight vector. Consider another dimension to represent the overall error. Plot the error on the weight space. Suppose it looks like the curve in Fig. 7.20. When we begin training, let us presume we are at point E1. As we reduce the error during the training, we arrive at point E2. If we use $\beta e(L, k)x(L, j)$ to calculate $\Delta w(L, k, j)$, then we may stop our training at E2, which we should not. E2 is a *local minimum:* there is another point, namely E4, where the error is less than the error at E2. Point E4 is the *global minimum:* there is no point at which the error is less than that at E4. Ideally, the training should stop at E4. While training, if we arrive at E2 from E1, we need some momentum to carry us past E2 to E3, from where we can then descend to E4. An analogy is how momentum keeps a sled going downhill even though the sled may pass over short flat stretches or bumps till the sled stops at the bottom of the hill: E2–E3 are the metaphoric bump and E4 is the bottom of the hill.

This momentum in training a neunet is provided by adding $\alpha \Delta w^{\mathrm{prev}}(L, k, j)$ to $\beta e(L, k)x(L, j)$, for calculating the value of $\Delta w(j)$. In $\alpha \Delta w^{\mathrm{prev}}(L, k, j)$, known as the *momentum term*, $0 \leq \alpha \leq 1$ is the *momentum constant* (also known as the 'inertia constant'), whose value we can select; and $\Delta w^{\mathrm{prev}}(L, k, j)$ is the value of $\Delta w(L, k, j)$ from the previous iteration, the most recent instance when $\Delta w(L, k, j)$ was evaluated—for the first iteration, the value of the momentum term is assumed to be zero. By using the *momentum term* with a non-zero value of α, we try to change $w(L, k, j)$ in the same direction as in the previous iteration; the momentum from the previous iteration influences the change in $w(L, k, j)$ in the current iteration. This attempts to change the weight in the same direction as in the previous iteration, thus preventing the weight from oscillating around a local minimum. This, in turn, speeds up the training. The momentum term is strong enough to push us past a local minimum, but not strong enough to push us past a global minimum. Accordingly, for $1 \leq L \leq G$, where $1 \leq k \leq N(L)$, and $B \leq j \leq [N(L-1)]$, we use the *training formula*

$$\Delta w(L, k, j) = \beta e(L, k) x(L, j) + \alpha \Delta w^{\text{prev}}(L, k, j) \qquad (7.42)$$

The *backpropagation procedure* to design and train a multilayer neunet is presented below. Explanations accompany the steps in the procedure. It will be easier to understand the backpropagation procedure if the procedure to train the single-neurode neunet is clear (described in Section 7.2), which can also be viewed as a simpler version of the procedure below (some duplication in the description of the two procedures is inevitable).

1. Fix the number G of layers to be used in the neunet. Researchers have argued that a two-layer neunet using the sigmoid activation function can solve any classification problem provided no limit is placed on the number of neurodes in the hidden layer. The neunet may accordingly need a large number of neurodes in the hidden layer. This does not mean that a two-layer neunet is the best. Neunets with more layers may be more efficient and effective but it is often recommended to begin with a few layers. If the neunet takes too long to train, or it does not recognize the training patterns well, then the number of layers can be increased to train the neunet.

2. Specify the number of non-biasing neurodes in each layer of the neunet. The stipulations of the classification problem for which the neunet is being designed will usually dictate the number of neurodes in layer zero and layer G: layer zero receives the attribute values of a given pattern, and layer G outputs values indicating the class of the pattern. A large number of attributes and how they are coded (an example on coding attributes is given in Section 7.7) for input could require a large number $N(0)$ of neurodes in layer zero. Similarly, a large number of classes and how they are coded for output could

require a large number $N(G)$ of neurodes in layer G. Decide the number $N(L)$ of neurodes[1] to be put in layers $1 \leq L < G$, that is, in the hidden layers. It has been found in practice that if too many neurodes are put in the hidden layers, then the neunet will usually recognize the training patterns well, but not so the recall patterns: the neunet can be said to have memorized the training patterns, or to have learned the training patterns by rote. Then again, too few neurodes in the hidden layers usually cause the neunet to over-generalize. Training the neunet may take long, and still the neunet may misclassify many recall patterns. A suggestion is to begin with few layers and few neurodes. Train the neunet. If it takes too long to train, or if it misrecognizes too many patterns, then increase the layers and the number of neurodes in each layer. The ideal is to have the fewest possible layers and neurodes for the neunet to do the job you want it to do. As is apparent from Eqn (7.36) and (7.37), the more layers and the more neurodes the neunet contains, the more weights the neunet will need to learn.

3. For each class, specify the desired output $z(G, k)$ for every neurode $u(G, k)$, for $1 \leq k \leq N(G)$. No two classes should have the same desired output.

4. Decide whether to use biasing neurodes in the neunet. If biasing neurodes are used, assign 0 to variable B; otherwise, assign 1 to B. Equation 7.38 shows how many more weights the neunet has to learn with biasing neurodes than without biasing neurodes. The more weights the neunet has to learn, the slower the training will be. A suggestion is to begin without biasing neurodes. If the neunet misrecognizes too many patterns, then add biasing neurodes to the neunet, and try to train the neunet again.

5. Select the activation function $\mathcal{A}$ for neurode $u(L, k)$, where $1 \leq L \leq G$ and $1 \leq k \leq N(L)$. The function should be monotonically increasing and differentiable. The sigmoid or the tanh function is recommended. It has been observed that the sigmoid function is selected more often in practice.

---

[1]The author is not aware of any mathematical results that can help the reader decide with certainty the number of neurodes in such layers.

6. Define observed output $y(G, k)$ to be some function of the activation value $a(G, k)$ of neurode $u(G, k)$, for $1 \leq k \leq N(G)$.

7. Decide whether case updating or epoch updating of the weights should be used. As mentioned in Section 7.2, in case updating, the weights are updated after processing every training pattern, whereas, in epoch updating, the weights are updated after processing all the training patterns.

8. Fix a heuristic value of the *learning constant* $0 < \beta \leq 1$. As mentioned in Section 7.2, a low value of β slows down the training, but a high value of β can cause the weights to oscillate. If training is to be done only once—typically for classification problems that do not change over time—then slow training may be considered to be acceptable. Often, β takes a value of 0.5. If, however, the training patterns are noisy, then a lower value of β may be more appropriate.

9. Choose a heuristic value for the momentum constant $0 \leq \alpha \leq 1$. Often, in practice, $\alpha = 0.9$.

10. Specify a heuristic value for epoch_bound to indicate within how many epochs the training of the neunet is expected to finish, that is, how many times each training pattern is expected to be processed.

11. Let $|V|$ be the number of training patterns in the training set V. It is desirable to have approximately equal number of patterns from each class; otherwise the neunet may learn frequent classes well, but the infrequent classes not so well. Moreover, it has been found in practice that it is better if the training patterns are fed (step 22) to the neunet in random order rather than all patterns of a class sequentially one after another. Since the patterns are fed to the neunet in the order in which they occur in the training set, the patterns should be placed in a random order in the training set.

12. If $B = 0$, then do
    for $L = 1, 2, \ldots, (G - 1)$ do
    $I(L, 0) \leftarrow 1$
    If biasing neurodes are present, then input to each of them is 1. This value will be needed in step 24 to calculate the error in layers 1 to $(G - 1)$.

13. For $L = 1, 2, \ldots, G$ do
    for $k = 1, 2, \ldots, N(L)$ do
    for $j = B, (B + 1), \ldots, N(L - 1)$ do steps 13.1 and 13.2.

13.1. Initialize $w(L, k, j)$ to a non-zero random real value.

These are the weights the neunet has to learn. The random values will typically be a mixture of positive and negative ones. Obtaining random values was discussed in Section 4.4. If all initial weights are identical, then it can happen that all weights change identically during the training. This can hamper the training of the neunet. So initial weights should not all be the same.

13.2. $\Delta w^{\mathrm{prev}}(L, k, j) \leftarrow 0$

Initialized to zero for the first iteration, $\Delta w^{\mathrm{prev}}(L, k, j)$ is used in step 25.1 to calculate $\Delta w(L, k, j)$.

14. epoch $\leftarrow 0$

Initialize variable epoch to zero so that it can count the epochs.

15. epoch $\leftarrow$ epoch $+1$

Increment the counter for epochs by 1.

16. If epoch $>$ epoch_bound, then go to step 30 to report failure in training the neunet.

17. For $L = 1, 2, \ldots, G$ do

for $k = 1, 2, \ldots, N(L)$ do

for $j = B, (B+1), \ldots, N(L-1)$ do

$s(L, k, j) \leftarrow w(L, k, j)$

The weights at the beginning of the epoch are saved in array s. If it is found in step 27 that the neunet was already trained, then the procedure will return with the weights as they were at the beginning of the epoch.

18. If epoch updating was chosen in step 7, then

for $L = 1, 2, \ldots, G$ do

for $k = 1, 2, \ldots, N(L)$ do

for $j = B, (B+1), \ldots, N(L-1)$ do

$q(L, k, j) \leftarrow 0$

Element $q(L, k, j)$ is initialized to zero here, so that, in steps 25.3 and 28, it can be used as an accumulator to sum up the amount by which $w(L, k, j)$ should be updated.

19. $T \leftarrow 0$

Initialize variable T to zero so that it can be used as a counter for the training patterns within each epoch.

20. $T \leftarrow T + 1$

Increment the counter for the training patterns by 1.

21. If $T > |V|$, that is, all training patterns have been processed in the current epoch, then go to step 27 and check whether the neunet had been trained by the end of the epoch.

22. Feed signals $x(0,1)$ to $x[0, N(0)]$ to the neunet, the signals representing the attribute values of the training pattern T. Invoke the forward propagation procedure given in Section 7.5 to obtain the values of $I(L, k)$, for $1 \leq L \leq G$, where $1 \leq k \leq N(L)$. Moreover, obtain the values of $y(G, 1)$, $y(G, 2)$, ..., $y[G, N(G)]$.

23. For $k = 1, 2, \ldots, N(G)$ do
$$e(G, k) \leftarrow \mathcal{A}'[I(G, k)] \times [z(G, k) - y(G, k)]$$
This is taken from Eqn (7.40), where $e(G, k)$ is the error in the output of neurode $u(G, k)$ in the output layer, and $\mathcal{A}'[I(G, k)]$ is the derivative of the activation function $\mathcal{A}$ at $I(G, k)$, the value input to $u(G, k)$. The term $[z(G, k) - y(G, k)]$ gives the difference between the desired output for the class of the pattern T and the observed output.

24. For $L = (G - 1)$, $(G - 2)$, ..., 1 do
for $j = B$, $(B + 1)$, ..., $N(L)$ do
$$e(L, j) \leftarrow \mathcal{A}'[I(L, j)] \sum_{k=1}^{N(L+1)} w(L + 1, k, j)\, e(L + 1, k)$$
This is taken from Eqn (7.41), where $e(L, j)$ is the error in the output of neurode $u(L, j)$ in a non-output layer. $\mathcal{A}'[I(L, j)]$ is the derivative of the activation function $\mathcal{A}$ at $I(L, j)$, the value input to $u(L, j)$.

25. For $L = 1, 2, \ldots, G$ do
for $k = 1, 2, \ldots, N(L)$ do
for $j = B$, $(B + 1)$, ..., $N(L - 1)$ do steps 25.1 to 25.4.

   25.1. $\Delta w(L, k, j) \leftarrow \beta e(L, k) x(L, j) + \alpha \Delta w^{\mathrm{prev}}(L, k, j)$
   This is taken from Eqn (7.42).

   25.2. If case updating was chosen in step 7, then do
   $w(L, k, j) \leftarrow w(L, k, j) + \Delta w(L, k, j)$
   Weight $w(L, k, j)$ is modified by $\Delta w(L, k, j)$.

   25.3. If epoch updating was chosen in step 7, then do
   $q(L, k, j) \leftarrow q(L, k, j) + \Delta w(L, k, j)$

In the current epoch, $q(L, k, j)$ accumulates the modifications required in $w(L, k, j)$.

25.4. $\Delta w^{\text{prev}}(L, k, j) \leftarrow \Delta w(L, k, j)$

Update $\Delta w^{\text{prev}}(L, k, j)$ so it can be used the next time $\Delta w(L, k, j)$ is calculated.

26. Go to step 20 to feed the neunet with the current epoch's next training pattern.

27. For at least p per cent of the training patterns in the last epoch, if the differences between each observed output $y(G, k)$ and its corresponding desired output $z(G, k)$, for $1 \leq k \leq N(G)$, was less than or equal to some heuristically chosen small value ε, say 0.05, then consider the training to be finished. If we choose p to be less than 100, then we are allowing $(100 - p)$ per cent of the training patterns to be misclassified by the trained neunet. If training is not finished, go to step 28; otherwise, it means that the weights at the beginning of the epoch were the ones to be learned by the neunet. If epoch updating was chosen in step 7, then the weights have not yet changed in the current epoch, hence go to step 31 to return with the weights. If, however, we have done case updating in step 25.2, then the weights have changed during the epoch, and we have to restore the weights as they were at the beginning of the epoch. We had saved the weights in array s in step 17. Consequently,
for $L = 1, 2, \ldots, G$ do
for $k = 1, 2, \ldots, N(L)$ do
for $j = B, (B + 1), \ldots, N(L - 1)$ do
$w(L, k, j) \leftarrow s(L, k, j)$
Go to step 31 to return with the weights.

28. If epoch updating was chosen in step 7, then
for $L = 1, 2, \ldots, G$ do
for $k = 1, 2, \ldots, N(L)$ do
for $j = B, (B + 1), \ldots, N(L - 1)$ do
$w(L, k, j) \leftarrow w(L, k, j) + q(L, k, j)$
After the epoch, the weights are updated by the accumulated modification required for all of the patterns in the training set.

29. Go to step 15 to begin a new epoch.

30.   Return with a message that the neunet failed to be trained within epoch_bound epochs. If so, you can try to train the neunet again by changing, for instance, one or more of the following:
- The number G of layers in step 1
- The number of non-biasing neurodes in one or more layers in step 2
- In step 4, if there were no biasing neurodes, then insert them; or if there were biasing neurodes, then remove them
- The activation function selected in step 5
- The definition of the output function in step 6
- The method of updating the weights chosen in step 7, case updating to epoch updating, or conversely
- The value of the learning rate β selected in step 8
- The value of the momentum rate α selected in step 9
- The value of the epoch_bound chosen in step 10
- The initial random values of the weights $w(L, k, j)$ in step 13.1
- The value of p in step 27
- The value of ϵ in step 27

31.   Return from the procedure with weights $w(L, k, j)$ for $1 \leq L \leq G$, where $1 \leq k \leq N(L)$ and $B \leq j \leq N(L-1)$. These are the weights learned by the neunet.

7.7 Professor–Student Neunet

Suppose we have to design a neunet and train it on the professor–student training set given in Fig. 1.1. Let the design be as follows:

- Let us have $G = 2$, because it has been observed in practice that most classification problems can be solved by a two-layer neunet. The neunet thus has layers 0, 1, and 2.
- The attributes are HABIT having two values (gabby and quiet), EATS having three values (baked, fried, and roasted), and FOOTWEAR having two values (clogs and sandals). We need to code these non-numeric values into numeric values, since the signals flowing through a neunet are numeric. Let us code these non-numeric values as shown in Fig. 7.21. The figure shows we shall have seven signals $x(0, 1)$ to $x(0, 7)$ flowing into the zeroth layer. Thus the input layer needs seven neurodes, that is, $N(0) = 7$.

Attribute	Value	Signals entering zeroth layer of neunet		
HABIT	= Gabby	$x(0,1) = 0$	$x(0,2) = 1$	
	= Quiet	$x(0,1) = 1$	$x(0,2) = 0$	
EATS	= Fried	$x(0,3) = 0$	$x(0,4) = 0$	$x(0,5) = 1$
	= Roasted	$x(0,3) = 0$	$x(0,4) = 1$	$x(0,5) = 0$
	= Baked	$x(0,3) = 1$	$x(0,4) = 0$	$x(0,5) = 0$
FOOTWEAR	= Clogs	$x(0,6) = 0$	$x(0,7) = 1$	
	= Sandals	$x(0,6) = 1$	$x(0,7) = 0$	

FIGURE 7.21 Numeric codes chosen for the non-numeric attribute values for use by a neunet to be trained on the professor–student training patterns given in Fig. 1.1.

- Layer 2, the output layer, needs only one neurode $u(2,1)$ to distinguish the class of professor from the class of student. Thus, $N(2) = 1$. Let us decide to have desired output $z(2,1) = 1$ for the class of professor, and $z(2,1) = 0$ for the class of student.

- Having decided on the number of neurodes in each of the input and output layers, we next need to decide on the number $N(1)$ of non-biasing neurodes in layer 1, the hidden layer. This number is a heuristic. Since too many neurodes in the hidden layer cause the neunet to memorize the training set, leading the neunet to recognize the training patterns well, but not so the recall patterns, let us choose a small value of $N(1)$, for instance, $N(1) = 2$.

- Equation (7.38) shows that, if we have biasing neurodes in the neunet, it has to learn more weights, thereby slowing the training. So a usual recommendation is to start without biasing neurodes. If we fail to train the neunet, we can insert biasing neurodes and then try to train the neunet. Accordingly, let us choose not to have any biasing neurodes. With all the decisions made till now, we can say that the neunet will look like the one shown in Fig. 7.22.

- The sigmoid function with constant $c = 1$, whose plot is shown in Fig. 7.18, is chosen to be the activation function. Thus, for $1 \leq L \leq 2$ and $1 \leq k \leq N(L)$, the activation value of neurode $u(L,k)$ with input

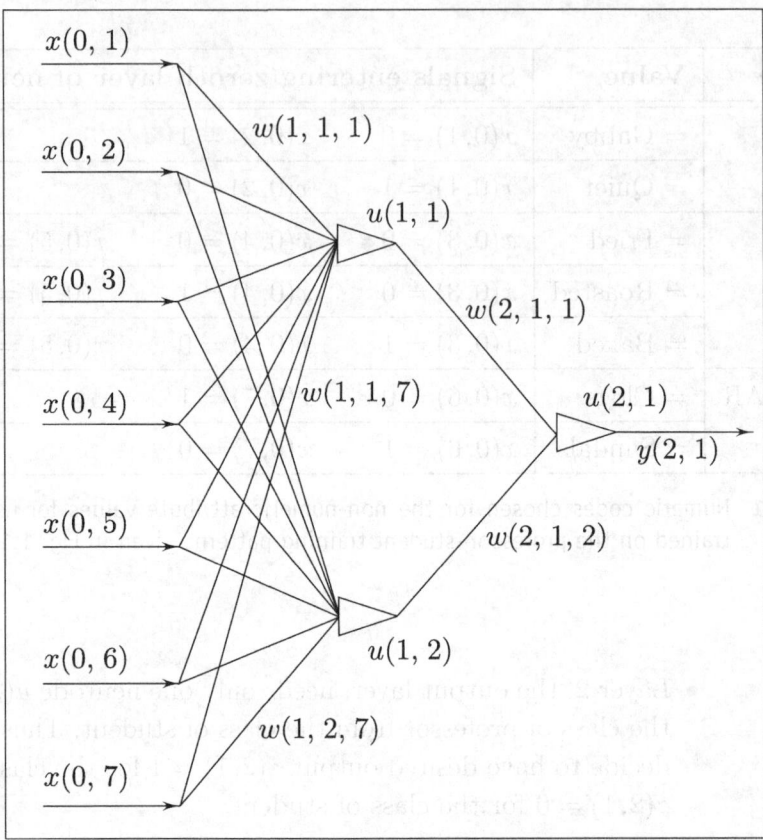

FIGURE 7.22 A neunet for recognizing the professor–student patterns given in Figs 1.1 and 1.2. The input layer has seven neurodes, the hidden layer has two neurodes, and the output layer has one neurode. No biasing neurodes are being used.

$I(L, k)$ is

$$a(L, k) = \frac{1}{1 + e^{-I(L,k)}}$$

- From neurode $u(2, 1)$, let us choose the observed output

$$y(2, 1) = \begin{cases} 0 & \text{if } a(G, k) < 0.05 \\ a(2, 1) & \text{if } 0.05 \le a(2, 1) \le 0.95 \\ 1 & \text{otherwise} \end{cases}$$

Coupling $y(2, 1)$ with the desired output $z(2, 1)$ described above, we can say that, for a given pattern, when $y(2, 1) = 0$, then the pattern is being recognized as a student, and when $y(2, 1) = 1$, then the pattern is being recognized as a professor.

- Since epoch updating is less likely to cause oscillations in weights than case updating, let us opt for epoch updating to train the neunet.
- Let us choose learning constant $\beta = 0.5$ and momentum constant $\alpha = 0.9$.
- Let epoch_bound = 40, expecting that the training will finish within that many epochs.
- We will consider the training finished, when for 100% of the training patterns, the observed output $y(2, 1)$ will be identical to the desired output $z(2, 1)$.

With the choices made above, let us invoke a random number generator (discussed in Section 4.4) to initialize weights $w(L, k, j)$ for $1 \leq L \leq 2$, where $1 \leq k \leq N(L)$, and $1 \leq j \leq N(L - 1)$. Let us suppose that the random initial values of the weights are as shown in the second column of Fig. 7.23. Then, if we are to implement the backpropagation training procedure of Section 7.6, we will notice that the training is finished in 11 epochs, and the weights become as shown in the third column of Fig. 7.23.

As an example, let us work through the forward propagation procedure of Section 7.5 to see how the signals would flow through the trained neunet for the training pattern T1 given in Fig. 1.1. It has these attribute values: HABIT equal to gabby, EATS equal to baked, and FOOTWEAR equal to clogs. According to Fig. 7.21, these values will be coded so that the following signals flow into the zeroth layer:

$$x(0, 1) = 0$$
$$x(0, 2) = 1$$
$$x(0, 3) = 1$$
$$x(0, 4) = 0$$
$$x(0, 5) = 0$$
$$x(0, 6) = 0$$
$$x(0, 7) = 1$$

Weight	Value after	
	Random initialization	Neunet's training
$w(1,1,1)$	-0.1352	-0.8368
$w(1,1,2)$	-0.1351	-0.9784
$w(1,1,3)$	-0.1949	-1.9110
$w(1,1,4)$	-0.0211	0.4040
$w(1,1,5)$	0.1647	-0.0889
$w(1,1,6)$	-0.0812	0.2308
$w(1,1,7)$	0.2428	-1.6140
$w(1,2,1)$	-0.1689	0.6690
$w(1,2,2)$	0.1917	0.0466
$w(1,2,3)$	0.1566	2.7410
$w(1,2,4)$	0.1102	-1.0610
$w(1,2,5)$	0.0469	-0.6733
$w(1,2,6)$	-0.1201	-1.6790
$w(1,2,7)$	0.1986	2.4510
$w(2,1,1)$	0.4979	18.3200
$w(2,1,2)$	0.1761	-7.0830

FIGURE 7.23 Weights of the connectors of the neunet given in Fig. 7.22 at the beginning of the training and after finishing the training.

Unchanged, the signals enter the first layer; hence $x(0,j)$ becomes $x(1,j)$, for $1 \leq j \leq 7$. Input to neurode $u(1,1)$ is

$$I(1,1) = \sum_{j=1}^{7} w(1,1,j) \, x(1,j) = 4.503$$

The values of the weights are taken from the third column of Fig. 7.23 and the results have been rounded off to three places after the decimal point.

Name of pattern	Signals $x(0,1)$ to $x(0,7)$	$I(1,1)$	$I(1,2)$	$a(1,1)$	$a(1,2)$	$I(2,1)$	$a(2,1)$	$y(2,1)$
T1	01 100 01	−4.503	5.239	0.011	0.995	−6.845	0.001	0
T2	01 010 10	−0.344	−1.726	0.415	0.151	6.532	0.999	1
T3	01 100 10	−2.659	1.109	0.066	0.752	−4.126	0.016	0
T4	10 001 10	−0.695	−1.683	0.333	0.157	4.990	0.993	1
T5	01 001 01	−2.681	1.824	0.064	0.861	−4.925	0.007	0
T6	10 100 10	−1.797	1.731	0.142	0.850	−3.412	0.032	0
T7	01 001 10	−0.837	−2.306	0.302	0.091	4.896	0.993	1
T8	10 001 01	−2.540	2.447	0.071	0.920	−5.212	0.005	0
R1	10 100 01	−4.362	5.861	0.013	0.997	−6.833	0.001	0
R2	10 010 10	−0.202	−2.071	0.450	0.112	7.446	0.999	1
R3	01 010 01	−2.118	1.437	0.107	0.808	−3.757	0.023	0
R4	10 010 01	−2.047	2.059	0.114	0.887	−4.186	0.015	0

FIGURE 7.24 Values propagated when attribute values of the professor–student training patterns T1 to T8 given in Fig. 1.1 and the recall patterns R1 to R4 given in Fig. 1.2 are fed to the neunet given in Fig. 7.22. To understand the entries in the second column, consider, for instance, pattern T1, whose attribute values are HABIT = gabby, EATS = baked, and FOOTWEAR = clogs. According to the numeric codes chosen in Fig. 7.21, these attribute values are represented by the string 01 100 01, the space within the string being shown only for your ease in reading. This means that the signals fed to the zeroth layer of the neunet have the following values: $x(0,1) = 0$, $x(0,2) = 1$, $x(0,3) = 1$, $x(0,4) = 0$, $x(0,5) = 0$, $x(0,6) = 0$, and $x(0,7) = 1$. If $y(2,1) = 0$, then the corresponding pattern is recognized as a student, and if $y(2,1) = 1$, then it is recognized as a professor. See Fig. 7.25.

Similarly, input to neurode $u(1,2)$ is

$$I(1,2) = \sum_{j=1}^{7} w(1,2,j)\, x(1,j) = 5.239$$

| NAME of | Attributes | | | Classi- |
pattern	HABIT	EATS	FOOTWEAR	fication
T1	Gabby	Baked	Clogs	Student
T2	Gabby	Roasted	Sandals	Professor
T3	Gabby	Baked	Sandals	Student
T4	Quiet	Fried	Sandals	Professor
T5	Gabby	Fried	Clogs	Student
T6	Quiet	Baked	Sandals	Student
T7	Gabby	Fried	Sandals	Professor
T8	Quiet	Fried	Clogs	Student
R1	Quiet	Baked	Clogs	Student
R2	Quiet	Roasted	Sandals	Professor
R3	Gabby	Roasted	Clogs	Student
R4	Quiet	Roasted	Clogs	Student

FIGURE 7.25 Classifications of the professor–student training patterns T1 to T8 given in Fig. 1.1 and the recall patterns R1 to R4 given in Fig. 1.2 by the trained neunet shown in Fig. 7.22, with the weights as shown in the third column of Fig. 7.23, and the propagated values as given in Fig. 7.24.

Then the activation value of neurode $u(1,1)$ is

$$a(1,1) = \frac{1}{1 + e^{-I(1,1)}} = 0.011$$

and the activation value of neurode $u(1,2)$ is

$$a(1,2) = \frac{1}{1 + e^{-I(1,2)}} = 0.995$$

Flowing to the second layer, $a(1,1)$ becomes $x(2,1)$, and $a(1,2)$ becomes $x(2,2)$. Then, input to neurode $u(2,1)$ is

$$I(2,1) = \sum_{j=1}^{2} w(2,1,j)\, x(2,j) = -6.845$$

and the neurode's activation value is

$$a(2,1) = \frac{1}{1 + e^{-I(2,1)}} = 0.001$$

Since this activation value is less than 0.05, then according to the definition of output chosen above,

$$y(2,1) = 0$$

According to the desired output values defined above for the two classes, we can say that the neunet has recognized pattern T1 as a student.

We can carry out the above calculations for the remaining training patterns given in Fig. 1.1 and for the recall patterns given in Fig. 1.2. The values of the signals propagated through the trained neunet for all these patterns are given in Fig. 7.24. The classifications of these patterns are then given in Fig. 7.25, which shows that the trained neunet recognized all training patterns correctly. Moreover, the classifications of the recall patterns are incidentally identical to the Naive Bayes classifications shown in Fig. 5.2, identical to the nearest neighbour classifications shown in Fig. 6.1, and identical to the classifications shown in Fig. 2.11 for the decision tree shown in Fig. 2.9, the tree obtained by maximizing the ratio of information gain.

7.8 Comments on Multilayer Neunets

It may be apparent by now that, in a neunet, every neurode is autonomous: there is no supervisor neurode. It is an established result that a neunet can do whatever a Turing machine—that is, any programmable machine—can do, but a neunet may not be always efficient. One would not want to design and train a neunet to solve partial differential equations. Such a neunet may require many layers and many neurodes, which may then take an impractically long time to train, since many epochs may be required to complete the training.

While learning to drive a motor car, we sometimes drive either to the right or the left instead of driving steadily. The car zigzags. We gradually learn to reduce the zigzag, until we can drive steadily. In a somewhat analogous way, a neunet learns to adjust its weights during training. Once we have learned to drive, we can drive instinctively. Similarly, once a neunet has learned its weights, it can quickly—as if by instinct—identify the class of the pattern whose attribute values are input to the neunet. A neunet does not, however, provide any explanation for its output, unlike, say, a decision tree (Chapters 2 and 3), where we can trace down the tree to see why a given pattern was classified the way it was. Humans, too, sometimes fail to provide a reasoning for their classifications: in Chapter 1, it was pointed out that although you may be able to distinguish a dog from a cat, you may have difficulty explaining what differentiates the two animals. We can say neunets model the brain. Methods such as decision trees model the mind. Humans need both a brain and a mind. Similarly, we need both neunet and non-neunet approaches to pattern recognition.

The idea of neurodes—small electronic or software-simulated processors— was inspired by neurons, which are small biological processors in the brain. A human brain, with a typical weight of 1.5 kg, is estimated to contain approximately 10^{11} (100 billion) neurons. The neurons receive, process, and transmit electrochemical signals while carrying messages in the brain. The interconnections between the neurons are far more complicated than the interconnection between the neurodes described in this chapter. A neuron in the brain may typically have 10^3 to 10^5 inputs, and similarly 10^3 to 10^5 outputs. Knowledge in the human brain is believed to be stored in neurons and their interconnections. The neurons together with their interconnections are estimated to have the ability to store 10^{14} (100 trillion) bits of information. A neunet can be constructed electronically. An electronic neunet carries signals at about two-thirds the speed of light, that is, 20 cm per nanosecond (10^{-9} seconds). Electrochemical signals in the human brain are a million times slower.

In practice, neunets are often simulated by software. The forward propagation (Section 7.5) and the backpropagation (Section 7.6) procedures do considerable matrix manipulation. To choose a programming language to simulate neunets, one may choose a language in which matrix manipulation can be coded easily.

Summary

Patterns from classes that are linearly separable can be recognized by single-layer neunets. Patterns from classes that are not linearly separable cannot be recognized by single-layer neunets, but can be recognized by multilayer neunets. The backpropagation procedure is commonly used to train multilayer neunets. The training may take long, but once a neunet is trained to recognize patterns from some given classes, it can quickly recognize recall patterns from those classes. It is, however, not able to explain its classifications.

Exercises

Before reading the exercises below, readers are advised to take a look at Fig. 7.16 to review the notation.

1. Write a computer program in any language of your choice to implement the backpropagation procedure of Section 7.6. The procedure invokes the forward propagation procedure of Section 7.5, hence that is also required to be coded. After a neunet has been trained, the forward propagation procedure can be used to classify a given training or recall pattern.

2. Design a two-layer neunet and use the program developed for the first exercise to train it to model an XOR gate (see Section 7.4), first without biasing neurodes, and then with biasing neurodes. The training set is shown in Fig. 7.9.

3. Design a two-layer neunet with an appropriate number of neurodes in the hidden layer, and use the program developed for the first exercise above to train the neunet on the professor–student training patterns given in Fig. 1.1. Do not insert any biasing neurodes. Let $N(0) = 7$ as in Fig. 7.22. Let $N(2) = 2$ with two neurodes $u(2, 1)$ and $u(2, 2)$. For the class professor, the desired outputs are $z(2, 1) = 1$ and $z(2, 2) = 0$, and for the class student $z(2, 1) = 0$ and $z(2, 2) = 1$. With the trained neunet, classify the patterns given in Fig. 1.1 and the recall patterns given in Fig. 1.2. Compare the results with those given in Fig. 7.25.

4. Repeat the third exercise after inserting biasing neurodes.

5. Suppose you are given this thumb rule:

   For large training sets, let the number of neurodes in the hidden layer be approximately between 50% to 150% of the number of training patterns.

Experiment with this rule on a small training set. Consider a neunet with $N(0) = 7$ and $N(2) = 1$, such as in Fig. 7.22. It has no biasing neurodes. For $N(1) = 4$, 5, ..., 12, use the program developed for the first exercise to train the neunet on the professor–student patterns given in Fig. 1.1. The attribute values and the output values may be coded as in Section 7.7. Test the neunet on the given training set and on the recall set given in Fig. 1.2. For each value of $N(1)$, comment on the number of epochs it took to train the neunet and on your classification results.

6. Assume attribute HABIT does not exist in the professor–student training set given in Fig. 1.1 and the recall set given in Fig. 1.2. Design a neunet and use the program developed in the first exercise above to train it on such a training set. Use the trained neunet to classify the training and the recall patterns. Are the classifications same as those in Fig. 7.25, in which HABIT is an attribute? If so, why is it so? (*Hint*: See Fig. 2.9.)

7. Read step 27 of the backpropagation procedure of Section 7.6. The step describes the criterion for finishing the training. An alternative criterion to finish training a neunet is the following:

> In the last epoch, if the proportional change in each of the weights $w(L, k, j)$, for $1 \leq L \leq G$, where $1 \leq k \leq N(L)$ and $B \leq j \leq N(L-1)$, was less than or equal to some heuristically chosen small value, say 0.05 (each of the weights changed by 5% or less), then the training of the neunet is considered to have been finished, and consequently return from the procedure with the weights $w(L, k, j)$. When changes in weights have become small, we are assuming that the observed outputs and the desired outputs of the neurodes in layer G are close enough, hence the training is finished.

Make changes in the program developed for the first exercise to incorporate this alternative. A user using this training program should be able to choose the criterion to finish training.

8. Repeat Exercises 2 to 6 so that the training finishes because of the criterion incorporated for the seventh exercise.

9. Exercises 2 to 6 have been done, using different criteria to finish training a neunet. Compare the results under the two criteria.

10. Use the program developed for the first exercise above (with possible changes made in the sixth exercise) to do the following and compare results with the corresponding parts of the first and second exercises of Chapter 6.

   (a) Train a neunet on the training set given in Fig. 6.2. Classify the training patterns and the recall patterns given in the same figure.

   (b) Train a neunet on the training patterns given in Fig. 6.3. Classify the training patterns and the recall patterns given in the same figure.

   (c) Train a neunet on the training set of computer science departments described in the tenth exercise of Chapter 2. Classify the training patterns and the corresponding recall patterns.

11. Section 3.1 described how decision trees are adapted to handle missing attribute values. Can neunets be adapted to handle missing attribute values? If not, why? If yes, describe the adaptations required. Remember, there may be missing attribute values in the training set. So, changes may be needed in the procedure of training the neunet. When there are missing attribute values in a recall pattern, changes may be required in the procedure of classification of the pattern.

Learning Objectives

This chapter contains the following topics:

- using linear classifiers when classes are linearly separable in the attribute space
- maximizing linear discriminant functions
- procedure to train a linear classifier
- the special case of two linearly separable classes
- training a linear classifier in a higher dimensional attribute space

8

Linear Classification

8.1 Training a Linear Classifier

A Naive Bayes classifier, with binary attributes, can be formulated as a linear classifier, as explained in Section 5.5. Similarly, Section 6.5 explains how a linear classifier can classify a recall pattern into the class of the nearest prototype. Moreover, when a neurode models an AND gate (Section 7.2), or it models an OR gate (Section 7.3), then it works as a linear classifier. Having seen these instances of linear classifiers, it should now be easier to understand and discuss linear classifiers in general.

For the training patterns belonging to $m \geq 1$ classes C_1 to C_m, let the attribute array $\bar{A} = A_1, A_2, \ldots, A_M$, for $M \geq 1$, contain numeric attribute values. Each training pattern can be represented as a point in the A_1-A_2-$\cdots$-A_M coordinate space. If we intuitively feel that the classes are linearly separable (that is, hyperplanes separate patterns of one class from another class), then we can try to develop a linear classifier. To do that, we establish m discriminant functions $g_1(\bar{A})$ to $g_m(\bar{A})$ (one for each class) as follows:

$$g_1(\bar{A}) = w_{10} + w_{11}A_1 + w_{12}A_2 + \cdots + w_{1M}A_M$$

$$g_2(\bar{A}) = w_{20} + w_{21}A_1 + w_{22}A_2 + \cdots + w_{2M}A_M$$

$$\vdots$$

$$g_m(\bar{A}) = w_{m0} + w_{m1}A_1 + w_{m2}A_2 + \cdots + w_{mM}A_M$$

The w's represent weights. If we succeed in finding the values of these weights so that for a pattern of class C_k, for $1 \leq k \leq m$, and with attribute array $\bar{A}$, the discriminant function $g_k(\bar{A})$ has the largest value among $g_1(\bar{A})$ to $g_m(\bar{A})$, then we have trained the linear classifier. The procedure to train the classifier is the following:

1. Choose the maximum number of iterations desired through the training set. Call this number the epoch_bound.

2. Choose a positive value for *learning constant* β, to be used in steps 8.1.3.1 and 8.1.3.2 to update the weights. A good choice is $0 < \beta \leq 1$ as advised in Section 7.2 (training a neurode).

3. Let $|V|$ be the number of training patterns in the training set V.

4. Let m be the number of classes C_1 to C_m in the training set.

5. Let M be the number of attributes.

6. $A_0 \leftarrow 1$.

   A_0 will be used in updating the weights in steps 8.1.3.1 and 8.1.3.2.

7. For $k = 1, 2, \ldots, m$ do

   for $i = 0, 1, \ldots, M$ do

   Initialize weight w_{ki} to a random value.

   These are the weights the procedure has to learn. Obtaining random values was discussed in Section 4.4.

8. For epoch $= 1, 2, \ldots,$ epoch_bound do steps 8.1 and 8.2.

   8.1. For $T = 1, 2, \ldots, |V|$ do steps 8.1.1 to 8.1.3.

   8.1.1. Let $\bar{A} = A_1, A_2, \ldots, A_M$ be the attribute array of the Tth training pattern.

   8.1.2. For $k = 1, 2, \ldots, m$ do

   $$g_k(\bar{A}) \leftarrow \sum_{i=0}^{M} w_{ki} A_i$$

   8.1.3. If the training pattern belongs to class C_k, but $g_j(\bar{A})$ is larger than $g_k(\bar{A})$, where $j \neq k$, then for $i = 0, 1, \ldots, M$ do steps 8.1.3.1 and 8.1.3.2.

   8.1.3.1. $w_{ki} \leftarrow w_{ki} + \beta A_i$

   8.1.3.2. $w_{ji} \leftarrow w_{ji} - \beta A_i$

   8.2. For at least p per cent of the training patterns, if no weights changed in the epoch, then the training is finished, hence go to step 10 to return with the weights. If we choose p to be less than 100, then we are allowing $(100 - p)$ per cent of the training patterns to be misclassified. Ideally, p should be 100.

9. Return with a message that training failed within the epoch_bound specified. The procedure to train the linear classifier can be repeated by choosing a larger epoch_bound in step 1. Alternatively, it can be concluded that it is not possible to build a linear classifier because the classes are not linearly separable in the attribute space.

10. Return with trained weights w_{ki}, for $1 \leq k \leq m$ and for $0 \leq i \leq M$. It is an established mathematical result that if the classes are indeed linearly separable, then the procedure will at last return with the trained weights.

It will not be surprising if similarities in the above procedure and the procedure to train a neurode given in Section 7.2 are observed.

After having trained the classifier, if while classifying a recall pattern later, it is found that two or more discriminant functions have the highest values, then the pattern can be rejected.

8.2 The Two-Class Case

If there are two linearly separable classes C_1 and C_2 in the attribute space, then the two linear discriminant functions, according to Section 8.1, will be

$$g_1(\bar{A}) = w_{10} + w_{11}A_1 + w_{12}A_2 + \cdots + w_{1M}A_M$$
$$g_2(\bar{A}) = w_{20} + w_{21}A_1 + w_{22}A_2 + \cdots + w_{2M}A_M$$

Classify a pattern in class C_1 if $g_1(\bar{A}) > g_2(\bar{A})$, classify it in class C_2 if $g_2(\bar{A}) > g_1(\bar{A})$, and reject it if $g_1(\bar{A}) = g_2(\bar{A})$. Let us subtract the second discriminant function from the first discriminant function to obtain

$$g_1(\bar{A}) - g_2(\bar{A}) = (w_{10} - w_{20}) + (w_{11} - w_{21})A_1 + \cdots + (w_{1M} - w_{2M})A_M$$

To simplify the notation, let us define $g(\bar{A}) = g_1(\bar{A}) - g_2(\bar{A})$ for the left-hand side of the equation. Similarly, let us define $w_i = (w_{1i} - w_{2i})$, for $0 \leq i \leq M$, for the right-hand side of the equation. Then the above equation can be written as

$$g(\bar{A}) = w_0 + w_1 A_1 + w_2 A_2 + \cdots + w_M A_M \tag{8.1}$$

On being given a pattern with attribute array $\bar{A}$, classify it in class C_1 if $g(\bar{A}) > 0$, classify it in class C_2 if $g(\bar{A}) < 0$, and reject it if $g(\bar{A}) = 0$. Instead of learning $2(M + 1)$ weights for $g_1(\bar{A})$ and $g_2(\bar{A})$, the classifier now needs to learn only $(M + 1)$ weights for $g(\bar{A})$. The neurode in Section 7.2 developed a single discriminant function for modelling the two-class AND gate, and similarly the neurode in Section 7.3 developed a single discriminant function for modelling the two-class OR gate. The procedure to learn the weights of a single linear discriminant function is the following:

1. Choose a value for epoch_bound, the maximum number of times iteration through the training set is desired.

2. Choose a positive value for *learning constant* β, to be used in steps 7.1.3 and 7.1.4 to update the weights. A good choice is $0 < \beta \leq 1$ as advised in Section 7.2 (training a neurode).

3. Let $|V|$ be the number of training patterns in the training set V.

4. Let M be the number of attributes.

5. $A_0 \leftarrow 1$.
   A_0 will be used in updating the weights in steps 7.1.3 and 7.1.4.

6. For $i = 0, 1, \ldots, M$ do
   Initialize weight w_i to a random value.
   These are the weights the procedure has to learn. Obtaining random values was discussed in Section 4.4.

7. For epoch $= 1, 2, \ldots$, epoch_bound do steps 7.1 and 7.2.

   7.1. For $T = 1, 2, \ldots, |V|$ do steps 7.1.1 to 7.1.4.

   7.1.1. Let $\bar{A} = A_1, A_2, \ldots, A_M$ be the attribute array of the Tth training pattern.

   7.1.2.
   $$g(\bar{A}) \leftarrow \sum_{i=0}^{M} w_i A_i$$
   This discriminant function is based on Eqn (8.1).

   7.1.3. If the training pattern belongs to class C_1, but is being classified into class C_2 because $g(\bar{A}) < 0$, then for $i = 0, 1, \ldots, M$ do $w_i \leftarrow w_i + \beta A_i$

   7.1.4. If the training pattern belongs to class C_2, but is being classified into class C_1 because $g(\bar{A}) > 0$, then for $i = 0, 1, \ldots, M$ do $w_i \leftarrow w_i - \beta A_i$

   7.2. For at least p per cent of the training patterns, if no weights changed in the epoch, then the training is finished, hence go to step 9 to return with the weights. If we choose p to be less than 100, then we are allowing $(100 - p)$ per cent of the training patterns to be misclassified. Ideally, p should be 100.

8. Return with a message that training failed within the epoch_bound specified. This procedure may be repeated by choosing a larger epoch_bound in step 1. Alternatively, it may be concluded that it is not possible to build a linear classifier because the two classes are not linearly separable in the attribute space.

9. Return with trained weights w_i, for $0 \leq i \leq M$. It is an established mathematical result that if the two classes are indeed linearly separable, then the procedure will at last return with the trained weights.

The above procedure is similar to the procedure described in Section 8.1. In fact, there is some repetition in the description, but that is unavoidable to prevent any confusion.

8.3 Higher-Dimensional Attribute Space

After repeated attempts of training a linear classifier, if the weights for the linear discriminant function are not found, it may be concluded that it is not possible to find the weights because the classes are not linearly separable in the given attribute space. The classifier can then be trained in a higher-dimensional space in which the classes are linearly separable. The discussion on how to do this can become abstract and difficult to understand. It is best understood with an example.

Figure 8.1 shows a training set with two attributes A_1 and A_2, and two classes C_1 and C_2. To train a linear classifier on this training set, we will first establish the linear discriminant function [based on Eqn (8.1)]

$$g(\bar{A}) = w_0 + w_1 A_1 + w_2 A_2 \tag{8.2}$$

We fail to learn the weights. Let us conclude that the classes are not linearly separable. The above discriminant function is of degree 1, for that is the highest power to which A_1 and A_2 are raised. It is possible that a quadratic

NAME of	Values of attributes		Class
pattern	A_1	A_2	
X1	-1	-1	C_1
X2	-1	1	C_2
X3	1	1	C_1
X4	1	-1	C_2

FIGURE 8.1   A specimen training set of two numeric attributes and two classes.

surface, represented by a function of degree 2, separates the two classes. So we increase the degree of the discriminant function by adding the terms of the right-hand side of the equation

$$(A_1 + A_2)^2 = A_1^2 + 2A_1A_2 + A_2^2$$

to Eqn (8.2). So the discriminant function becomes

$$g(\bar{A}) = w_0 + w_1A_1 + w_2A_2 + w_3A_1^2 + w_4A_1A_2 + w_5A_2^2$$

The three terms added have been associated with weights w_3, w_4, and w_5. Let us assume, we have a revised set of attributes A_1' to A_5' such that

$$A_1' = A_1$$
$$A_2' = A_2$$
$$A_3' = A_1^2$$
$$A_4' = A_1A_2$$
$$A_5' = A_2^2$$

The training set with the revised attributes is shown in Fig. 8.2. The discriminant function becomes

$$g(\bar{A}) = w_0 + w_1A_1' + w_2A_2' + w_3A_3' + w_4A_4' + w_5A_5'$$

If we train the linear classifier in the A_1'-A_2'-A_3'-A_4'-A_5' coordinate space, we will obtain the discriminant function

$$g(\bar{A}) = A_4'$$
$$= A_1A_2$$

If the attributes of a pattern make $g(\bar{A})$ positive, we classify the pattern into class C_1; if, however, the attributes make it negative, we classify the pattern into class C_2. The classifier learned the weights of the discriminant function using revised attributes as an intermediate step. In the A_1-A_2 coordinate space, the classes are not linearly separable, but they are linearly separable in the higher dimensional A_1'-A_2'-A_3'-A_4'-A_5' coordinate space.

In the above example, there are only four patterns, and if we are to represent them in the A_1-A_2 coordinate space, we can clearly see that the classes are not linearly separable. In general, there may be many patterns, the patterns having M (a large number) attributes, and it may be difficult to

Name of pattern	Values of revised attributes					
	A'_1 $=A_1$	A'_2 $=A_2$	A'_3 $=A_1^2$	A'_4 $=A_1 A_2$	A'_5 $=A_2^2$	Class
X1	-1	-1	1	1	1	C_1
X2	-1	1	1	-1	1	C_2
X3	1	1	1	1	1	C_1
X4	1	-1	1	-1	1	C_2

FIGURE 8.2 The training set given in Fig. 8.1 with the revised attributes A'_1 to A'_5.

visualize that the classes are not linearly separable in the M-dimensional coordinate space. So, when we repeatedly fail to train the linear classifier, we can conclude that the classes are not linearly separable. We can, of course, be wrong: had we tried longer by iterating more times through the training set, we might have succeeded (see epoch_bound in step 1 of the training procedures described in Sections 8.1 and 8.2). After we have failed to learn the weights, we increase the degree of the discriminant function by one, and try to train again. The function is no longer linear, but we can consider it to be linear in a higher-dimensional space, and try to train the classifier in this space. If we fail, we can again increase the degree of the discriminant function by one, that is, we can go to higher and higher dimensional space, until we either succeed, or give up. In increasing the degree of the discriminant function from $(n-1)$ to n, where $n \geq 2$, we add the terms that appear after expanding the M-attribute expression

$$(A_1 + A_2 + \cdots + A_M)^n$$

to the discriminant function, where M is the number of attributes. In the higher-dimensional space, patterns closest to the hyperplane separating the classes are known as *support vectors*. The farther away the support vectors are from the separating hyperplane, the farther away are the classes from one another, and thus the lower are the chances of later misclassifying recall patterns.

Summary

When classes are linearly separable in the attribute space, then a linear classifier can be trained. The classifier learns the weights of as many linear discriminant functions as there are classes. For the special two-class case, the amount of learning can be reduced so that the classifier learns the weights of only one linear discriminant function. If the classes are not linearly separable, the classifier may be trained in a higher-dimensional attribute space.

Exercises

1. Write a program in any chosen language to implement a linear classifier. It should learn the weights of as many discriminant functions as there are classes, but when there are two classes, it should learn the weights of only one discriminant function. If the classifier fails to be trained as a linear classifier, then it tries to be trained in a higher-dimensional attribute space. Use the program to do the following and compare the classification results obtained with those in the corresponding parts of the first and second exercises of Chapter 6, and in the corresponding parts of the tenth exercise of Chapter 7.

    (a) Train the classifier on the training set given in Fig. 6.2. Classify the training and recall patterns given in the same figure.

    (b) Train the classifier on the training set given in Fig. 6.3. Classify the training and recall patterns given in the same figure.

    If the classifier fails to be trained for either or both of the above training sets, try to explain why it failed.

Learning Objectives

This chapter contains the following topics:

- cross validation to estimate how a classifier will perform in general
- removing redundant attributes to speed up classification
- cross validation to select attributes that are useful for classification
- example of selecting attributes for the professor–student pattern of the first chapter

9

Cross Validation and Attribute Selection

9.1   Cross Validation Procedure

Suppose we have a training set V of cardinality $|V|$ and a classifier, for example, the nearest neighbour classifier of Chapter 6. We want to empirically estimate how well the classifier will perform in general. One way is to train the classifier on set V, and then testing it on V itself, observe how many patterns are classified correctly, how many are rejected (failed to classify), and how many are misclassified. Estimating the general performance of a classifier this way is questionable. Since it was trained on set V, its performance on V is expected to be much better than what it would perhaps be on some recall set. We can, however, estimate with some selected level of confidence how well the classifier will perform on a recall set, as discussed in Section 3.3. An alternative procedure, often adopted, to estimate a classifier's general performance is *cross validation* as described below:

1. Split training set V into mutually exclusive subsets $H_1, H_2, \ldots, H_q$, where $q \geq 1$, that is

$$V = \bigcup_{i=1}^{q} H_i$$

As far as possible, choose q and H_i such that the class distribution in each H_i is similar to the class distribution in V, and the cardinalities of the H_i's are equal, or approximately equal, to one another. Each H_i can contain one pattern, but the fewer the patterns in each H_i, the more computation required in steps that follow. In practice, the cardinality of each H_i is often taken to be about one-tenth the cardinality of the training set V.

2. For $i = 1, 2, \ldots, q$ do
   Train the classifier on the truncated training set $V - H_i$. Test it on H_i, as if H_i were a recall set. Since the class of every pattern in H_i is known, note the number of patterns in H_i that are classified correctly, the number rejected, and the number misclassified.

3. Sum for all H_i each of the number of patterns correctly classified, the number rejected, and the number misclassified. Divide each sum by $|V|$ to obtain the corresponding recognition rate, rejection rate, and error rate—the sum of the three rates should be equal to one. These

rates can give you an idea about how well the classifier will perform in general. These rates are based on training the classifier on truncated training sets. When we train the classifier on the full training set, it should perform better because the full training set is larger than each of the truncated training sets.

Suppose we have a training set V with two classifiers and we want to find out which one is expected to perform better so that we can choose it for an application. We can carry out the cross validation procedure with each classifier, and then compare their recognition, rejection, and error rates.

We will need to establish a criterion for comparing performance. The criterion of choosing the classifier with the lower error rate is debatable, for the classifier may also have a low recognition rate because of a high rejection rate. A suggested criterion is to prefer the classifier with the higher recognition rate, unless there are other reasons for not doing so, for instance, the classifier with the higher recognition rate may require so much computation, that it may not be worthwhile to adopt it for the marginal increase in recognition rate it provides. If both the classifiers have equal recognition rates, opt for the one with the lower error rate—this classifier will have a higher rejection rate, but it is usually better to have a classifier that rejects a pattern than misclassifies it.

9.2 Attribute Selection Procedure

The professor–student training patterns given in Fig. 1.1 have three attributes: HABIT, EATS, and FOOTWEAR. Section 2.5 uses ratio of information gain to build the decision tree given in Fig. 2.9. The tree examines only EATS and FOOTWEAR: it finds HABIT redundant, that is, not needed for classification.

If redundant attributes are used in other classifiers, say Bayes or nearest neighbour (Chapters 5 and 6), then the computation required for classification increases, thus slowing down the speed of classification. Suppose the values of $D \geq 1$ attributes $A_1, A_2, \ldots, A_D$ for the $|V|$ patterns of the training set V are given. Our objective is to find the attributes that are not useful for classification so that they can be discarded, and only $M \leq D$ attributes be retained. The time spent on selecting the M attributes should later turn out to be beneficial: with fewer attributes, classification should become faster.

The exhaustive technique to try out all the possible 2^D subsets of the D attributes for determining the subset that provides the best classification may require an impractically large amount of time, especially when D is large. To meet our objective, a more efficient way is to use the following *attribute selection* procedure, which invokes cross validation.

1. Select the classifier to be employed.

2. Decide on the criterion to be used to assess the classifier's performance, based on its recognition, rejection, and error rates, as discussed in Section 9.1.

3. Repeatedly invoke the cross validation procedure to run the classifier on the training set, the classifier using only one feature at a time. Thus, the classifier uses only attribute A_1 first, then A_2, and so on till A_D. For each attribute, record the recognition, rejection, and error rates. Based on the criterion established in step 2, select attribute A_i, for $1 \leq i \leq D$, with which the classifier gives the best performance.

4. Carry out cross validation with the classifier using two features at a time: $A_i A_1$, $A_i A_2$, ..., $A_i A_{i-1}$, $A_i A_{i+1}$, ..., $A_i A_D$. Select the pair with which the classifier gives the best performance. Suppose the best pair is $A_i A_j$, for $1 \leq i \leq D$, and $1 \leq j \leq D$.

5. Carry out cross validation with the classifier using three attributes at a time: $A_i A_j A_1$, $A_i A_j A_1$, and so on.

6. Thus at every step one attribute is added to the list of selected attributes. Proceed this way till the performance of the classifier is satisfactory. Select the attributes for which this happens and discard the remaining attributes.

To illustrate, let us apply the Naive Bayes classifier [Eqn (5.2)] on the professor–student training patterns given in Fig. 1.1 to select attributes from the given attributes HABIT, EATS, and FOOTWEAR. As is mentioned in Section 5.2 of Chapter 5, in practice, Bayesian estimates of probabilities are more useful than maximum-likelihood estimates. So let us use Bayesian estimates: prior probabilities from Eqn (5.4), and conditional probabilities from Eqn (5.6).

Figure 9.1 shows that, when we use only attribute EATS, four patterns are classified correctly and four misclassified. We can continue similarly one by

Applying the Naive Bayes classifier using attribute EATS to train on the professor–student patterns T5 to T8 given in Fig. 1.1, and test on the patterns T1 to T4. The Bayesian probability estimates obtained from T5 to T8 are

Probability	Bayesian estimates
$P(\text{professor})$	1/3
$P[(\text{EATS} = \text{baked})|\text{professor}]$	1/4
$P[(\text{EATS} = \text{fried})|\text{professor}]$	1/2
$P[(\text{EATS} = \text{roasted})|\text{professor}]$	1/4
$P(\text{student})$	2/3
$P[(\text{EATS} = \text{baked})|\text{student}]$	1/3
$P[(\text{EATS} = \text{fried})|\text{student}]$	1/2
$P[(\text{EATS} = \text{roasted})|\text{student}]$	1/6

We note that pattern T1 has EATS = baked; hence, to classify it, we compare $P(\text{professor}) \times P[(\text{EATS} = \text{baked}) \mid \text{professor}] = 1/3 \times 1/4 = 1/12$ with $P(\text{student}) \times P[(\text{EATS} = \text{baked}) \mid \text{student}] = 2/3 \times 1/3 = 2/9$. Since the latter is larger, we classify pattern T1 as a student, which is correct because that is its class given in Fig. 1.1. So T1 has been classified correctly. Proceeding this way, we will finally observe that two patterns (T1 and T3) are classified correctly, and two patterns (T2 and T4) are misclassified. Now we need to train the classifier on the patterns T1 to T4 and test on the patterns T5 to T8. The Bayesian probability estimates obtained from T1 to T4 are

Probability	Bayesian estimates
$P(\text{professor})$	1/2
$P[(\text{EATS} = \text{baked})|\text{professor}]$	1/5
$P[(\text{EATS} = \text{fried})|\text{professor}]$	2/5
$P[(\text{EATS} = \text{roasted})|\text{professor}]$	2/5
$P(\text{student})$	1/2
$P[(\text{EATS} = \text{baked})|\text{student}]$	3/5
$P[(\text{EATS} = \text{fried})|\text{student}]$	1/5
$P[(\text{EATS} = \text{roasted})|\text{student}]$	1/5

With the above probability estimates, two patterns (T6 and T7) are classified correctly and two patterns (T5 and T8) are misclassified. Overall, with EATS as the sole attribute, four patterns are classified correctly and four are misclassified.

FIGURE 9.1   Applying cross validation with the Naive Bayes classifier using attribute EATS on the professor–student patterns given in Fig. 1.1.

one for the other attributes. On doing so, it is observed that, with attribute HABIT, three patterns are classified correctly and five misclassified. With attribute FOOTWEAR, five patterns are classified correctly and three misclassified. Figure 9.1 shows only the calculations for EATS as the

Applying the Naive Bayes classifier using attributes FOOTWEAR and EATS to train on the professor–student patterns T5 to T8 given in Fig. 1.1, and test on the patterns T1 to T4. The Bayesian probability estimates obtained from T5 to T8 are

Probability	Bayesian estimates
$P(\text{professor})$	1/3
$P[(\text{EATS} = \text{baked})|\text{professor}]$	1/4
$P[(\text{EATS} = \text{fried})|\text{professor}]$	1/4
$P[(\text{EATS} = \text{roasted})|\text{professor}]$	1/2
$P[(\text{FOOTWEAR} = \text{clogs})|\text{professor}]$	1/3
$P[(\text{FOOTWEAR} = \text{sandals})|\text{professor}]$	2/3
$P(\text{student})$	2/3
$P[(\text{EATS} = \text{baked})|\text{student}]$	1/3
$P[(\text{EATS} = \text{fried})|\text{student}]$	1/2
$P[(\text{EATS} = \text{roasted})|\text{student}]$	1/6
$P[(\text{FOOTWEAR} = \text{clogs})|\text{student}]$	3/5
$P[(\text{FOOTWEAR} = \text{sandals})|\text{student}]$	2/5

We note that pattern T1 has EATS = baked and FOOTWEAR = clogs; hence, to classify it, we compare $P(\text{professor}) \times P[(\text{EATS} = \text{baked})|\text{professor}] \times P[(\text{FOOTWEAR} = \text{clogs}) |\text{professor}] = 1/3 \times 1/4 \times 1/3 = 1/36$ with $P(\text{student}) \times P[(\text{EATS} = \text{baked})|\text{student}] \times P[(\text{FOOTWEAR} = \text{clogs})|\text{student}] = 2/3 \times 1/3 \times 3/5 = 6/45$. Since the latter is larger, we classify pattern T1 as a student, which is correct because that is its class given in Fig. 1.1. So T1 has been classified correctly. Proceeding this way, we will finally observe that all four patterns T1 to T4 are classified correctly. Now we need to train the classifier on the patterns T1 to T4, and test on the patterns T5 to T8. Proceeding as above, we will observe that two patterns (T6 and T7) will be classified correctly and two patterns (T5 and T8) will be rejected. Overall, with FOOTWEAR and EATS as attributes, six patterns are classified correctly and two are rejected.

FIGURE 9.2 Applying cross validation with the Naive Bayes classifier using attributes EATS and FOOTWEAR on the professor–student patterns given in Fig. 1.1.

Attributes used	Patterns		
	Classified correctly	Rejected	Misclassified
EATS	4	0	4
HABIT	3	0	5
FOOTWEAR	5	0	3
FOOTWEAR, EATS	6	2	0
FOOTWEAR, HABIT	4	2	2

FIGURE 9.3   Recapitulation of the results of applying the attribute selection procedure to the professor–student patterns given in Fig. 1.1.

calculations for HABIT and FOOTWEAR proceed along the lines of EATS. Since FOOTWEAR classifies maximum number of patterns correctly, we select it.

We next take the two attributes FOOTWEAR and EATS. Figure 9.2 shows that, with these two attributes, six patterns are classified correctly and two are rejected. Now we take attributes FOOTWEAR and HABIT. With these two attributes, four patterns are classified correctly, two are rejected, and two are misclassified (calculations not shown for brevity). Since FOOTWEAR and EATS classify more patterns correctly than FOOTWEAR and HABIT, we select FOOTWEAR and EATS.

Since readers might want to review the classification results for each of the attributes, and for their pairs, these results have been recapitulated in Fig. 9.3.

The attribute selection procedure finds attributes FOOTWEAR and EATS more useful than attribute HABIT. This is coincidentally the same result[1] as that shown in Section 2.5, where the ratio of information gain is used to select attributes for the decision tree shown in Fig. 2.9.

---

[1]The author is not aware of any proof which says that the attribute selection procedure and the ratio of information gain will always select the same attributes. This is a topic for research.

Summary

The cross validation procedure can be used to assess the performance of a classifier. Furthermore, given a number of patterns with their attribute values, cross validation can be used to select the attributes that are the most useful for classifying the patterns.

Exercises

1. Manually apply the attribute selection procedure explained in Section 9.2 with the nearest neighbour classifier using Hamming distance on the professor–student patterns given in Fig. 1.1 to select attributes from their given attributes HABIT, EATS, and FOOTWEAR. Are the results obtained same as in Fig. 9.3, obtained by using the Naive Bayes classifier? Discuss the differences and similarities.

2. Write a program in a language of your choice to implement the attribute selection procedure. It should invoke the classifier that is being used; for example, the Naive Bayes classifier coded in the fifth exercise of Chapter 5, or the nearest neighbour classifier coded in the first exercise of Chapter 6.

3. Readers were asked to do the first exercise in this chapter manually. Repeat the same exercise using the program developed in the second exercise above.

Learning Objectives

This chapter contains the following topics:

- the underlying idea of clustering
- clustering as a form of unsupervised learning
- agglomerative hierarchical clustering: simple-linkage, farthest-linkage, complete-linkage, and centroid-linkage
- k-means clustering
- example of clustering the patterns in the professor–student training set of the first chapter

10

Clustering

10.1 Clustering is Grouping

If we group classification procedures based on whether they can generate prules after their training, then SpecToGen, GenToSpec, decision tree, and evolutionary procedures (Chapters 1 to 4) fall into the group of generating prules. Bayes, nearest neighbour, and neural nets (Chapters 5 to 7) fall into the group of not generating prules. The generation of prules serves as an attribute based on which we have carried out the grouping: members of a group are similar for that attribute. If, however, we group the classification procedures based on whether the procedure can be said to model the brain or whether they can be said to model the mind, then neural nets fall into the group modelling the brain, and the remaining procedures fall into the group modelling the mind. Grouping can be carried out based on more than one attribute. If we group the classification procedures on whether they generate prules after training as well as whether they model the brain or the mind, then we get three groups: (a) SpecToGen, GenToSpec, decision tree, and evolutionary procedures which model the mind and generate prules; (b) Bayes and nearest neighbour procedures which model the mind and do not generate prules; and (c) neural net which models the brain but does not generate prules.

Patterns can also be put into groups based on the values of their attributes. Such groups are called *clusters,* and the process of forming clusters is called *clustering.* The process has often been found to be useful in the exploratory stages of researching a domain to learn how patterns in the domain can be clustered. Clustering is also known as 'unsupervised learning', that is, there

NAME of pattern	Values of attributes	
	A_1	A_2
X1	1	1
X2	2	3
X3	3	1
X4	4	4
X5	5	2

FIGURE 10.1 Set of five patterns that are to be clustered.

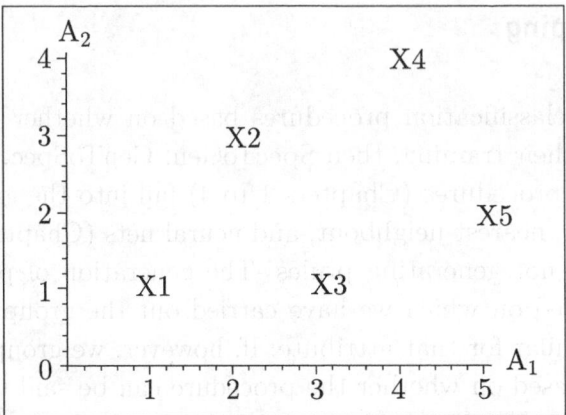

FIGURE 10.2 Representing each pattern given in Fig. 10.1 as a point in the A_1-A_2 coordinate space.

is no training set available to supervise, or guide, the learning. After we have clustered some given patterns, we may consider each cluster to be a class.

Study the five patterns X1 to X5 and the values of their numeric attributes A_1 and A_2 in Fig. 10.1. It is not necessary to know what the attributes mean; only their values need to be considered. Intuitively put the five patterns into two clusters such that the patterns within a cluster are more similar to one another than to patterns in another cluster. It might help to first represent each pattern as a point in the A_1-A_2 coordinate space as shown in Fig. 10.2; points in a cluster are closer to one another than to points in another cluster. Once you have finished clustering, ask a friend to form the clusters. It is possible that the clusters made by your friend are different from the clusters made by you.

10.2 Agglomerative Hierarchical Clustering

There exist procedures which—given $r \geq 1$ patterns and the values of their attributes $A_1, A_2, \ldots, A_M$—can put the patterns into $k \geq 1$ clusters, where $k \leq r$. Different procedures may, however, produce different clusters. A commonly adopted procedure has the following steps.

1. Choose a value for k, which denotes the number of clusters desired.
2. Consider each of the r patterns to be in a separate cluster. In other words, each cluster contains one pattern.

3. If there are k clusters, then return from the procedure with the k clusters.

4. Find the pairwise distances between the clusters. Merge the two closest clusters into one cluster, thus reducing the number of clusters by one. If there is more than one pair of clusters that are the closest to each other, then arbitrarily select a pair to merge, for instance, the first such pair encountered while scanning the clusters in some order. Go to step 3.

Step 4 above measures the distance between clusters. Before deciding how to measure that, we need to first decide how to measure the distance between patterns. Euclidean, city-block, Hamming distance (explained in Sections 6.2 and 6.3), or some function of these can be chosen to measure the distance. In practice, the Euclidean distance is the one most often chosen. Figure 10.3 shows the Euclidean distances between the patterns given in Fig. 10.1, which can be confirmed from Fig. 10.2.

The distances between two clusters C' and C'' can then be measured as follows. First calculate the pairwise distance for every pattern in cluster C' with every pattern in cluster C''. Then, the distance between cluster C' and cluster C'' can be defined in the following manner.

- The lowest of the pairwise distances—if this definition is chosen for clustering, then the procedure is known as *single-linkage*, or *nearest neighbour*, clustering.

Patterns	X1	X2	X3	X4	X5
X1	0	2.2	2	4.2	4.1
X2	2.2	0	2.2	2.2	3.2
X3	2	2.2	0	3.2	2.2
X4	4.2	2.2	3.2	0	2.2
X5	4.1	3.2	2.2	2.2	0

FIGURE 10.3 Euclidean distances between the five patterns given in Fig. 10.1. The distance between patterns X1 and X2 is $\sqrt{(2-1)^2 + (3-1)^2} = 2.2$ (rounded off). The diagonal values are zero because the distance between a pattern and itself is zero. The table is symmetric because, for example, the distance between patterns X1 and X2 is equal to the distance between X2 and X1.

To create two clusters of the five patterns X1 to X5 given in Fig. 10.1 by the **single-linkage** procedure using Euclidean distance, we begin by creating 5 clusters {X1}, {X2}, {X3}, {X4}, and {X5}. Considering the off-diagonal values in Fig. 10.3, we see that the minimum distance between any two clusters is 2: it is between {X1} and {X3}, which we merge into one cluster {X1, X3}. We then obtain the distances between the four clusters {X1, X3}, {X2}, {X4}, and {X5}, as given in the following symmetric table:

Clusters	{X1, X3}	{X2}	{X4}	{X5}
{X1, X3}	0	2.2	3.2	2.2
{X2}	2.2	0	2.2	3.2
{X4}	3.2	2.2	0	2.2
{X5}	2.2	3.2	2.2	0

Since we are using the single-linkage procedure, the distance between {X1, X3} and, say, {X4} is the lesser of the distance between X1 and X4, and that between X3 and X4, the two distances obtained from Fig. 10.3. Therefore, the distance between {X1, X3} and {X4} is $\min(4.2, 3.2) = 3.2$, which is what is shown in the table above. The other values in the table are obtained similarly. The smallest off-diagonal value in the table is 2.2. We can merge any pair of clusters having a distance of 2.2 between them. Let us select the pair corresponding to 2.2 we encounter first when we scan the table row by row. So we merge {X2} with {X1, X3}. We now have the three clusters {X1, X2, X3}, {X4}, and {X5}, the distances between which are in the following symmetric table:

Clusters	{X1, X2, X3}	{X4}	{X5}
{X1, X2, X3}	0	2.2	2.2
{X4}	2.2	0	2.2
{X5}	2.2	2.2	0

With our using the single-linkage procedure, the distance between {X1, X2, X3} and, say, {X4} is the least of the distances between X4 and either of X1, X2, and X3 (see Fig. 10.3). The distance between {X1, X2, X3} and {X4} accordingly is $\min(4.2, 2.2, 3.2) = 2.2$, as shown in the above table. The smallest off-diagonal value in the table is 2.2. Selecting the pair of clusters corresponding to 2.2 we encounter first when we scan the table row by row, we merge {X1, X2, X3} with {X4} to finally obtain these two clusters: {X1, X2, X3, X4} and {X5}.

FIGURE 10.4 An example of clustering the patterns X1 to X5 given in Fig. 10.1 by the single-linkage procedure using Euclidean distance. In the tables, the diagonal elements are zero because the distance between a cluster and itself is zero. The above tables can be compared with those given in Figs 10.5 to 10.7.

To create two clusters of the five patterns X1 to X5 given in Fig. 10.1 by the **complete-linkage** procedure using Euclidean distance, we begin by creating 5 clusters {X1}, {X2}, {X3}, {X4}, and {X5}. Considering the off-diagonal values given in Fig. 10.3, we see that the minimum distance between any two clusters is 2: it is between {X1} and {X3}, which we merge into one cluster {X1, X3}. We then obtain the distances between the four clusters {X1, X3}, {X2}, {X4}, and {X5}, as given in the following symmetric table:

Clusters	{X1,X3}	{X2}	{X4}	{X5}
{X1,X3}	0	2.2	4.2	4.1
{X2}	2.2	0	2.2	3.2
{X4}	4.2	2.2	0	2.2
{X5}	4.1	3.2	2.2	0

Since we are using the complete-linkage procedure, the distance between {X1, X3} and, say, {X4} is the greater of the distance between X1 and X4, and that between X3 and X4, the two distances obtained from Fig. 10.3. Therefore, the distance between {X1, X3} and {X4} is max(4.2, 3.2) = 4.2, which is what is shown in the table above. The other values in the table are obtained similarly. The smallest off-diagonal value in the table is 2.2. We can merge any pair of clusters having a distance of 2.2 between them. Let us select the pair corresponding to 2.2 we encounter first when we scan the table row by row. So we merge {X2} with {X1, X3}. We now have the three clusters {X1, X2, X3}, {X4}, and {X5}, the distances between which are in the following symmetric table:

Clusters	{X1, X2, X3}	{X4}	{X5}
{X1, X2, X3}	0	4.2	4.1
{X4}	4.2	0	2.2
{X5}	4.1	2.2	0

With our using the complete-linkage procedure, the distance between {X1, X2, X3} and, say, {X4} is the greatest of the distances between X4 and either of X1, X2, and X3 (see Fig. 10.3). The distance between {X1, X2, X3} and {X4} accordingly is max(4.2, 2.2, 3.2) = 4.2, as shown in the above table. The smallest off-diagonal value in the table is 2.2. We merge {X5} with {X4} to finally obtain these two clusters: {X1, X2, X3} and {X4, X5}.

FIGURE 10.5 An example of clustering the patterns X1 to X5 given in Fig. 10.1 by the complete-linkage procedure using Euclidean distance. In the tables, the diagonal elements are zero because the distance between a cluster and itself is zero. The above tables can be compared with those given in Figs 10.4, 10.6, and 10.7.

To create two clusters of the five patterns X1 to X5 given in Fig. 10.1 by the **average-linkage** procedure using Euclidean distance, we begin by creating 5 clusters {X1}, {X2}, {X3}, {X4}, and {X5}. Considering the off-diagonal values given in Fig. 10.3, we see that the minimum distance between any two clusters is 2: it is between {X1} and {X3}, which we merge into one cluster {X1, X3}. We then obtain the distances between the four clusters {X1, X3}, {X2}, {X4}, and {X5}, as given in the following symmetric table:

Clusters	{X1, X3}	{X2}	{X4}	{X5}
{X1, X3}	0	2.2	3.7	3.2
{X2}	2.2	0	2.2	3.2
{X4}	3.7	2.2	0	2.2
{X5}	3.2	3.2	2.2	0

Since we are using the average-linkage procedure, the distance between {X1, X3} and, say, {X4} is the average of the distance between X1 and X4, and that between X3 and X4, the last two distances obtained from Fig. 10.3. Therefore, the distance between {X1, X3} and {X4} is $(4.2 + 3.2)/2 = 3.7$, which is what is shown in the table above. The other values in the table are obtained similarly. For simplicity, the values have been rounded off to one place after the decimal. The smallest off-diagonal value in the table is 2.2. We can merge any pair of clusters having a distance of 2.2 between them. Let us select the pair corresponding to 2.2 we encounter first when we scan the table row by row. So we merge {X2} with {X1, X3}. We now have the three clusters {X1, X2, X3}, {X4}, and {X5}, the distances between which are in the following symmetric table:

Clusters	{X1, X2, X3}	{X4}	{X5}
{X1, X2, X3}	0	3.2	3.2
{X4}	3.2	0	2.2
{X5}	3.2	2.2	0

With our using the average-linkage procedure, the distance between {X1, X2, X3} and, say, {X4} is the average of the distances between X4 and each of X1, X2, and X3 (see Fig. 10.3). The distance between {X1, X2, X3} and {X4} accordingly is $(4.2 + 3.2 + 2.2)/3 = 3.2$, as shown in the above table. The smallest off-diagonal value in the table is 2.2. We merge {X5} with {X4} to finally obtain these two clusters: {X1, X2, X3} and {X4, X5}.

FIGURE 10.6   An example of clustering the patterns X1 to X5 given in Fig. 10.1 by the average-linkage procedure using Euclidean distance. In the tables, the diagonal elements are zero because the distance between a cluster and itself is zero. The above tables can be compared with those given in Figs 10.4, 10.5, and 10.7.

- The highest of the pairwise distances—if this definition is chosen, then the procedure is known as *complete-linkage*, or *farthest neighbour*, clustering.

- The average of the pairwise distances—if this definition is chosen, then the procedure is known as *average-linkage* clustering.

In addition to this, there is one more way to measure the distance between clusters. The distance between clusters C' and C'' is defined to be equal to the distance between their centroids. For $1 \leq i \leq m$, the value of the A_ith coordinate of the centroid of a cluster is the mean of the values of attribute A_i of the patterns in the cluster (this is what you did for a prototype in Section 6.5). If this definition is chosen to measure the distance between clusters, then the clustering procedure is known as *centroid-linkage* clustering.

As examples, the patterns given in Fig. 10.1 are clustered using Euclidean distance by single-linkage clustering in Fig. 10.4, by complete-linkage clustering in Fig. 10.5, by average-linkage clustering in Fig. 10.6, and by centroid-linkage clustering in Fig. 10.7. The four procedures do not produce identical clusters. For calculating the distance between clusters, the single-linkage and complete-linkage clustering procedures are dependent on single patterns in clusters. An outlier (defined in Section 6.1) in either of the clusters can affect the distance measured between the two clusters. Accordingly, it is recommended that the average-linkage or the centroid-linkage clustering procedures be preferred over the single-linkage and complete-linkage clustering procedures.

All the above four procedures are said to be *agglomerative hierarchical* clustering procedures. They are agglomerative because they build larger clusters from smaller clusters, and they are hierarchical because the larger clusters are built by merging the smaller clusters.

10.3   *k*-means Clustering

A non-hierarchical way of clustering is the *k-means* procedure, which is as follows:

1.   Consider k of the r given patterns to each form a cluster. We thus have k clusters, each with one pattern. The remaining patterns remain as

To create two clusters of the five patterns X1 to X5 given in Fig. 10.1 by the **centroid-linkage** procedure using Euclidean distance, we begin by creating 5 clusters {X1}, {X2}, {X3}, {X4}, and {X5}. Considering the off-diagonal values given in Fig. 10.3, we see that the minimum distance between any two clusters is 2: it is between {X1} and {X3}, which we merge into one cluster. The coordinates of the centroid of the resulting cluster {X1, X3} are $[(1 + 3)/2 = 2, (1 + 1)/2 = 1]$. We then obtain the distances between the four clusters {X1, X3}, {X2}, {X4}, and {X5}, as given in the following symmetric table:

Clusters	{X1, X3}	{X2}	{X4}	{X5}
{X1, X3}	0	2	3.6	3.2
{X2}	2	0	2.2	3.2
{X4}	3.6	2.2	0	2.2
{X5}	3.2	3.2	2.2	0

Since we are using the centroid-linkage procedure, the distance between {X1, X3} and, say, {X4} is $\sqrt{(4 - 2)^2 + (4 - 1)^2} = 3.6$, which is what is shown in the table above. The other values in the table are obtained similarly. For simplicity, the values have been rounded off to one place after the decimal. The smallest off-diagonal value in the table is 2; it is between {X1, X3} and {X2}, which we merge into one cluster. The coordinates of the centroid of the resulting cluster {X1, X2, X3} are $[(1 + 2 + 3)/3 = 2, (1 + 3 + 1)/3 = 1.67]$. We now have the three clusters {X1, X2, X3}, {X4}, and {X5}, the distances between which are in the following symmetric table:

Clusters	{X1, X2, X3}	{X4}	{X5}
{X1, X2, X3}	0	3.1	3
{X4}	3.1	0	2.2
{X5}	3	2.2	0

With our using the centroid-linkage procedure, the distance between {X1, X2, X3} and, say, {X4} is $\sqrt{(4 - 2)^2 + (4 - 1.67)^2} = 3.1$, as shown in the above table. The smallest off-diagonal value in the table is 2.2. We merge {X5} with {X4} to finally obtain these two clusters: {X1, X2, X3} and {X4, X5}.

FIGURE 10.7 An example of clustering the patterns X1 to X5 given in Fig. 10.1 by the centroid-linkage procedure using Euclidean distance. In the tables, the diagonal elements are zero because the distance between a cluster and itself is zero. The above tables can be compared with those given in Figs 10.4 to 10.6.

such. The k patterns to form clusters may be chosen arbitrarily: they are usually the first k patterns. In this step, the centroid of a given cluster is the coordinates of the pattern in the cluster in the A_1-A_2- $\cdots$ -A_M coordinate space.

2. For $i = k+1,\ k+2,\ \ldots,\ r$ do
   Put the ith pattern in the cluster whose centroid is the nearest to the pattern. Compute the cluster's new centroid.

3. For $i = 1,\ 2,\ \ldots,\ r$ do steps 3.1 to 3.3.

   3.1. Let $\mathcal{C}'$ be the cluster which has the ith pattern. Calculate the distance between the ith pattern and the centroids of each of the k clusters. If the pattern is closest to the centroid of $\mathcal{C}'$, then the pattern need not change its cluster, hence go to step 3.3.

   3.2. Let $\mathcal{C}''$ be the cluster whose centroid is the closest to the ith pattern. Move the pattern from cluster $\mathcal{C}'$ to cluster $\mathcal{C}''$. Compute the new centroids of clusters $\mathcal{C}'$ and $\mathcal{C}''$. Go to step 3.

   3.3. Continue.

4. No patterns changed clusters in the last iteration; hence return from the procedure with the k clusters.

As should be clear from above, the k-means procedure begins by putting each pattern in one of the k clusters (steps 1 and 2). Then, in step 3, it ensures that every pattern is in a cluster whose centroid is closest to it: if moving a pattern from one cluster to another is required, the procedure does so. This makes the k-means procedure preferable to the rest of the clustering procedures. The procedure is said to be *partitional* because it partitions the set of patterns into clusters.

As an example, let us create two clusters of the five patterns X1 to X5 given in Fig. 10.1 by the k-means procedure using Euclidean distance. As mentioned in step 1 of the k-means procedure, we begin by considering each of the first two patterns to form their own clusters. From Fig. 10.1, we see that the coordinates of the centroids of the two clusters, and the coordinates of the three remaining patterns are

$$\begin{array}{ccccc} \{X1\} & \{X2\} & X3 & X4 & X5 \\ (1,\,1) & (2,\,3) & (3,\,1) & (4,\,4) & (5,\,2) \end{array}$$

We are now in step 2 of the k-means procedure. The distance between pattern X3 and the centroid of cluster {X1} is

$$\sqrt{(3-1)^2 + (1-1)^2} = 2$$

The distance between pattern X3 and the centroid of cluster {X2} is

$$\sqrt{(3-2)^2 + (1-3)^2} = 2.2$$

X3 is closer to the centroid of {X1} than to {X2}; hence we put X3 into {X1}. The coordinates of the centroid of the resulting cluster {X1, X3} become $[(1+3)/2 = 2, (1+1)/2 = 1]$. Then the coordinates of the centroids of the two clusters, and the coordinates of the two remaining patterns are

$$\{X1, X3\} \quad \{X2\} \quad X4 \quad X5$$
$$(2,\ 1) \quad (2,\ 3) \quad (4,\ 4) \quad (5,\ 2)$$

The distance between pattern X4 and the centroid of cluster {X1, X3} is

$$\sqrt{(4-2)^2 + (4-1)^2} = 3.6$$

The distance between pattern X4 and the centroid of cluster {X2} is

$$\sqrt{(4-2)^2 + (4-3)^2} = 2.2$$

X4 is closer to the centroid of {X2} than to {X1, X3}, hence we put X4 into {X2}. The coordinates of the centroid of the resulting cluster {X2, X4} become $[(2+4)/2 = 3, (3+4)/2 = 3.5]$. The coordinates of the centroids of the two clusters, and the coordinates of the remaining pattern are then

$$\{X1, X3\} \quad \{X2, X4\} \quad X5$$
$$(2,\ 1) \quad (3,\ 3.5) \quad (5,\ 2)$$

The distance between pattern X5 and the centroid of cluster {X1, X3} is

$$\sqrt{(5-2)^2 + (2-1)^2} = 3.2$$

The distance between pattern X5 and the centroid of cluster {X2, X4} is

$$\sqrt{(5-3)^2 + (2-3.5)^2} = 2.5$$

X5 is closer to the centroid of {X2, X4} than to {X1, X3}, hence we put X5 into {X2, X4}. The coordinates of the centroid of the resulting cluster {X2, X4, X5} become $[(2 + 4 + 5)/3 = 3.67, (3 + 4 + 2)/3 = 3]$. The coordinates of the centroids of the two clusters are accordingly

$$\{X1, X3\} \quad \{X2, X4, X5\}$$
$$(2, 1) \quad\quad (3.67, 3)$$

Each of the five patterns is now in a cluster. We are now in step 3 of the *k*-means procedure. We next need to iterate through the patterns and move each pattern from the cluster it is in to the cluster whose centroid is nearest to the pattern. Accordingly, we notice that the distance between X1 and the centroid of cluster {X1, X3} is

$$\sqrt{(1 - 2)^2 + (1 - 1)^2} = 1$$

The distance between X1 and the centroid of cluster {X2, X4, X5} is

$$\sqrt{(1 - 3.67)^2 + (1 - 3)^2} = 3.3$$

X1 is closer to the centroid of {X1, X3} than to {X2, X4, X5}; hence X1 stays in the cluster it is in. We next notice that the distance between X2 and the centroid of cluster {X1, X3} is

$$\sqrt{(2 - 2)^2 + (3 - 1)^2} = 2$$

The distance between X2 and the centroid of cluster {X2, X4, X5} is

$$\sqrt{(2 - 3.67)^2 + (3 - 3)^2} = 1.67$$

X2 is closer to the centroid of {X2, X4, X5}, than to {X1, X3}; hence X2 stays in the cluster it is in. Proceeding along, we notice that the distance between X3 and the centroid of cluster {X1, X3} is

$$\sqrt{(3 - 2)^2 + (1 - 1)^2} = 1$$

The distance between X3 and the centroid of cluster {X2, X4, X5} is

$$\sqrt{(3 - 3.67)^2 + (1 - 3)^2} = 2.1$$

X3 is closer to the centroid of {X1, X3} than to {X2, X4, X5}; hence X3 stays in the cluster it is in. In the same vein, we notice that the distance between X4 and the centroid of cluster {X1, X3} is

$$\sqrt{(4-2)^2 + (4-1)^2} = 3.6$$

The distance between X4 and the centroid of cluster {X2, X4, X5} is

$$\sqrt{(4-3.67)^2 + (4-3)^2} = 1.1$$

X4 is closer to the centroid of {X2, X4, X5} than to {X1, X3}; hence X4 stays in the cluster it is in. Finally, we notice that the distance between X5 and the centroid of cluster {X1, X3} is

$$\sqrt{(5-2)^2 + (2-1)^2} = 3.2$$

The distance between X5 and the centroid of cluster {X2, X4, X5} is

$$\sqrt{(5-3.67)^2 + (2-3)^2} = 1.7$$

Procedure	The two clusters obtained
Single-linkage	{X1, X2, X3, X4} and {X5}
Complete-linkage	
Average-linkage	
Centroid-linkage	{X1, X2, X3} and {X4, X5}
k-means	{X1, X3} and {X2, X4, X5}

FIGURE 10.8 Recapitulating the results of clustering the five patterns given in Fig. 10.1 into two clusters by different procedures, using Euclidean distance. Not all procedures give the same final clusters. The results of the different linkage procedures have been obtained from Figs 10.4 to 10.7. The results of the k-means procedure are from the example solved in Section 10.3. The single-linkage, average-linkage, and centroid-linkage procedures have coincidentally produced identical clusters[1].

---

[1]The author is not aware of any theory that says the three procedures will always produce identical clusters from a given set of patterns.

X5 is closer to the centroid of {X2, X4, X5} than to {X1, X3}; hence X5 stays in the cluster it is in.

We have completed an iteration, and no pattern changed its cluster; hence we terminate with the clusters as they exist: {X1, X3} and {X2, X4, X5}. Had any pattern changed its cluster, we would need to recalculate the centroids of the newly formed clusters and iterate again. For the sake of brevity, this example terminates after one iteration. In practice, many such iterations may be required.

The clusters resulting from the patterns given in Fig. 10.1, using the five clustering procedures discussed above: single-linkage, complete-linkage, average-linkage, centroid-linkage, and k-means are recapitulated in Fig. 10.8 for the convenience of the readers.

In calculating Euclidean distances while clustering, finding the square roots can be omitted to save computation. It will not affect the results, square root being monotonic. Nonetheless, in the preceding examples, the square roots have been calculated to avoid any confusion in understanding the procedures.

10.4 Non-numeric Attributes

Clustering is mostly carried out on patterns with numeric attributes. With non-numeric attributes, it is difficult to decide into which cluster a pattern should be put, because it is difficult to measure the similarity between patterns. Nonetheless, clustering can be carried out by using a measure such as Hamming distance. As mentioned in Section 6.3, the Hamming distance between two patterns is the number of attributes in which the two patterns have different values. Figure 10.9 shows the clustering of the professor–student patterns given in Fig. 1.1 by the average-linkage procedure using Hamming distance. The figure shows that the two clusters obtained are {T1, T2, T3, T4, T6, T7} and {T5, T8}. It is debatable whether the two clusters are meaningful. They are not the same as the classes of the patterns.

To create two clusters of the eight patterns T1 to T8 given in Fig. 1.1, which have non-numeric attribute values, by the average-linkage procedure using Hamming distance, we begin by creating eight clusters {T1}, {T2}, {T3}, {T4}, {T5}, {T6}, {T7}, {T8}. The Hamming distances between the eight clusters is given by the following table:

Patterns	{T1}	{T2}	{T3}	{T4}	{T5}	{T6}	{T7}	{T8}
{T1}	0	2	1	3	1	2	2	2
{T2}	2	0	1	2	2	2	1	3
{T3}	1	1	0	2	2	1	1	3
{T4}	3	2	2	0	2	1	1	1
{T5}	1	2	2	2	0	3	1	1
{T6}	2	2	1	1	3	0	2	2
{T7}	2	1	1	1	1	2	0	2
{T8}	2	3	3	1	1	2	2	0

The smallest off-diagonal value in the preceding table is 1, and scanning the table row by row, we encounter this value for the first time between {T1} and {T3}. So, we merge these two clusters. The distances between the seven clusters we now have are shown in the following table:

Patterns	{T1,T3}	{T2}	{T4}	{T5}	{T6}	{T7}	{T8}
{T1, T3}	0	1.5	2.5	1.5	1.5	1.5	2.5
{T2}	1.5	0	2	2	2	1	3
{T4}	2.5	2	0	2	1	1	1
{T5}	1.5	2	2	0	3	1	1
{T6}	1.5	2	1	3	0	2	2
{T7}	1.5	1	1	1	2	0	2
{T8}	2.5	3	1	1	2	2	0

The smallest off-diagonal value in the preceding table is 1, and scanning the table row by row, we encounter this value for the first time between {T2} and {T7}. So, we merge these two clusters. The distances between the six clusters we now have are shown in the following table:

(contd)

(*contd*)

Patterns	{T1, T3}	{T2, T7}	{T4}	{T5}	{T6}	{T8}
{T1, T3}	0	1.5	2.5	1.5	1.5	2.5
{T2, T7}	1.5	0	1.5	1.5	2	2.5
{T4}	2.5	1.5	0	2	1	1
{T5}	1.5	1.5	2	0	3	1
{T6}	1.5	2	1	3	0	2
{T8}	2.5	2.5	1	1	2	0

The smallest off-diagonal value in the preceding table is 1, and scanning the table row by row, we encounter this value for the first time between {T4} and {T6}. So, we merge the two clusters. The distances between the five clusters we now have are shown in the following table:

Patterns	{T1, T3}	{T2, T7}	{T4, T6}	{T5}	{T8}
{T1, T3}	0	1.5	2	1.5	2.5
{T2, T7}	1.5	0	1.75	1.5	2.5
{T4, T6}	2	1.75	0	2.5	1.5
{T5}	1.5	1.5	2.5	0	1
{T8}	2.5	2.5	1.5	1	0

The smallest off-diagonal value in the preceding table is 1, between {T5} and {T8}. So, we merge the two clusters. The distances between the four clusters we now have are shown in the following table:

Patterns	{T1, T3}	{T2, T7}	{T4, T6}	{T5, T8}
{T1, T3}	0	1.5	2	2
{T2, T7}	1.5	0	1.75	2
{T4, T6}	2	1.75	0	2
{T5, T8}	2	2	2	0

The smallest off-diagonal value in the preceding table is 1.5, between {T1, T3} and {T2, T7}. So, we merge the two clusters. The distances between the three clusters we now have are shown in the following table:

Patterns	{T1, T2, T3, T7}	{T4, T6}	{T5, T8}
{T1, T2, T3, T7}	0	1.9	2
{T4, T6}	1.9	0	2
{T5, T8}	2	2	0

The smallest off-diagonal value in the preceding table is 1.9, between {T1, T2, T3, T7} and {T4, T6}. So, we merge them to finally obtain two clusters: {T1, T2, T3, T4, T6, T7} and {T5, T8}.

FIGURE 10.9 Clustering of the eight professor–student patterns T1 to T8 given in Fig. 1.1 by the average-linkage procedure using Hamming distance.

Summary

Clustering is a grouping of similar patterns. Five clustering procedures have been described: single-linkage, complete-linkage, average-linkage, centroid-linkage, and k-means. Clustering is best done on numeric attributes because with non-numeric attributes it is difficult to define the similarity of patterns.

Exercises

1. Figure 10.3 shows the Euclidean distances between the patterns given in Fig. 10.1. Build a table showing the city-block distances (defined in Section 6.2) between the same patterns. Using these distances, manually produce two clusters of the patterns by each of these procedures: (a) single-linkage, (b) complete-linkage, (c) average-linkage, (d) centroid-linkage, and (e) k-means. Compare the results with one another and with those of Fig. 10.8.

2. Write a program in a language of your choice to implement the agglomerative hierarchical clustering described in Section 10.2. The user should be able to choose the distance measure (Euclid or city-block) and the kind of linkage (single, complete, average, or centroid). Run the program to produce two clusters of the patterns given in Fig. 10.1 by each of these procedures: (a) single-linkage, (b) complete-linkage, (c) average-linkage, (d) centroid-linkage. Cluster first by using the Euclidean distance and then by the city-block distance.

3. Write a program in a language of your choice to implement k-means clustering. The user should be able to choose Euclid or city-block distance. Run the program to produce two clusters of the patterns given in Fig. 10.1. Cluster first by using the Euclidean distance and then by the city-block distance.

4. Using Hamming distance, manually create two clusters of the eight professor–student patterns T1 to T8 given in Fig. 1.1 by (a) single-linkage procedure and (b) complete-linkage procedure.

5. Using the programs developed in the second and third exercises above, experiment with different clustering procedures and distance measures to do the following.

    (a) Allocate the training patterns given in Fig. 6.2 to three clusters. Are the three classes given in the figure and the three clusters created using different procedures identical? Give reasons for your answers.

(b) Allocate the training patterns given in Fig. 6.3 to two clusters. Are the two classes as given in the figure and the two clusters created using different procedures identical? Give reasons for your answer.

(c) Allocate the twenty departments (see the tenth exercise of Chapter 2), to two clusters. Two classes for these departments had been created. Are the two classes and the two clusters created using the different procedures identical? Give reasons for your answer.

11

Syntactic Pattern Recognition

11.1   Strings and Grammars

Using rules of English grammar, we can recognize the following pattern as a sentence.

The cat sat on the mat.

Each word in the sentence can be viewed as a symbol. When patterns are represented as sequences of symbols, their recognition consists of determining whether a given grammar produced them.

An *alphabet* is a finite set of symbols. A *string* is a single symbol or a concatenated sequence of two or more symbols drawn from a given alphabet. Suppose $\mathcal{A}$ is an alphabet such that $\mathcal{A} = \{S,\ B,\ C,\ a,\ b,\ c\}$. Then,

$$a,\ aabc,\ BCbacC,\ bbbcccBBCb$$

are examples of strings drawn from $\mathcal{A}$.

The *length* of a string γ is the number of symbols in it: if $\gamma = bbbcccBBCb$ then the length of γ, denoted by $|\gamma|$, is 10. For ease in reading, a superscript with a symbol indicates the number of consecutive appearances of the symbol in a string: we can thus alternatively say $\gamma = b^3 c^3 B^2 Cb$.

In addition to the strings drawn from a given alphabet, there is ε, the empty string; its length being 0. Concatenating any string with ε leaves the string unchanged. Thus,

$$\gamma\varepsilon = \varepsilon\gamma = \gamma$$

A *grammar* is defined over a given alphabet. In the alphabet, it reserves a special symbol, usually S by convention, called the *start symbol*. The grammar also has one or more *rewriting rules* of the form $\alpha \longrightarrow \beta$, where α and β are strings such that $|\alpha| \geq 1$: this means that if there is a non-empty string α, it can be rewritten as β; in other words, α can be replaced by β. Since the rewriting rules help produce strings from the grammar, they are also known as *production rules*, (abbreviated as *prules*). Consider a grammar defined over alphabet $\mathcal{A} = \{S,\ B,\ C,\ a,\ b,\ c\}$, with prules

1. $S \longrightarrow CB$
2. $C \longrightarrow a$

3. $C \longrightarrow b$

4. $B \longrightarrow bc$

Then, we can produce the following string from the grammar:

$$S \overset{1}{\Longrightarrow} CB$$
$$\overset{2}{\Longrightarrow} aB$$
$$\overset{4}{\Longrightarrow} abc$$

S being the start symbol, we begin from S to produce a string. To avoid any confusion between a prule's definition and its application, or firing, a singly lined '$\longrightarrow$' is used to define the prule, and a doubly-lined '$\Longrightarrow$' to show its firing. The number above '$\Longrightarrow$' identifies the prule fired. Another string that can be produced from the above grammar is

$$S \overset{1}{\Longrightarrow} CB$$
$$\overset{3}{\Longrightarrow} bB$$
$$\overset{4}{\Longrightarrow} bbc$$

No other string can be produced from the grammar. The set $\{abc, bbc\}$ is said to be the *language* produced by the grammar. The symbols a, b, c are said to be *terminals*: the language produced comprises strings of terminals. Strings that belong to a language are said to be *sentences* in the language. The symbols S, B, C are said to be *non-terminals*: they are used in intermediate stages of producing the strings constituting the language. Unless otherwise mentioned, lower case English letters will be used for terminals, upper case English letters will be used for non-terminals, and lower case Greek letters will be used for strings containing only terminals, only non-terminals, or a mixture of both.

The alphabet $\mathcal{A}$ over which a grammar is defined is thus split into two disjoint subsets: $\mathcal{A}_T$ contains the terminals and $\mathcal{A}_N$ contains the non-terminals. The start symbol is contained in $\mathcal{A}_N$. In general, to define a grammar, we need to define these four items: $\mathcal{A}_N$, $\mathcal{A}_T$, the prules, and the start symbol, which here is reserved to be S, where S belongs to $\mathcal{A}_N$.

The set of all possible strings that can be drawn from $\mathcal{A}_T$ is denoted by $\mathcal{A}_T^+$. Since, there is no limit on the length of the strings that can be drawn from

$\mathcal{A}_T$, set $\mathcal{A}_T^+$ is infinite. Adding the empty string ε to the set $\mathcal{A}_T^+$ gives us the set $\mathcal{A}_T^*$, that is,

$$\mathcal{A}_T^* = \mathcal{A}_T^+ \cup \{\varepsilon\}$$

The language produced by a grammar is a subset of $\mathcal{A}_T^*$. There are four types of grammars depending on the constraints put on the prules of the form $\alpha \longrightarrow \beta$, where α contains at least one non-terminal.

- *Type 0* or *Free* or *Unrestricted* No restrictions are put on the prules. These grammars have not been found useful in practice.
- *Type 1* or *Context-Sensitive* In a context-sensitive grammar, in every prule

$$|\alpha| \leq |\beta|$$

Since $|\alpha| \geq 1$ and $|\varepsilon| = 0$, there cannot be a prule of the form $\alpha \longrightarrow \varepsilon$. With $\alpha = \alpha_1 I \alpha_2$ and $\beta = \alpha_1 \beta_1 \alpha_2$, a typical prule is of the form

$$\alpha_1 I \alpha_2 \longrightarrow \alpha_1 \beta_1 \alpha_2$$

in which β_1 can replace the non-terminal I, when I appears in the context of α_1 and α_2. Of course, both α_1 and α_2, or either of them, may be empty.

- *Type 2* or *context-free* As for a context-sensitive grammar, in every prule

$$|\alpha| \leq |\beta|$$

In a context-free grammar, α must be a single non-terminal, for example $I \longrightarrow \beta$. Thus, a string can replace non-terminal I free of the context in which I occurs.

- *Type 3* or *finite-state* or *regular* As for a context-free grammar

$$|\alpha| \leq |\beta|$$

where α is a single non-terminal. Moreover, in a finite-state grammar, β can either be a terminal or a terminal concatenated on its right to a non-terminal; for example,

$$I \longrightarrow a$$

or

$$I \longrightarrow aJ$$

Because of the way more and more constraints are put as we move from Type 0 to Type 3 grammars, we can say that, for $1 \leq i \leq 3$, every Type i grammar is a Type $(i - 1)$, but not conversely. Context-free grammars are the ones most often used in practice.

11.2 Chomsky Normal Form

Prules in a context-free grammar, which are each of the form $I \longrightarrow \beta$, can be transformed so that each prule then has the form

$$I \longrightarrow AB$$

or

$$I \longrightarrow a$$

that is, in every prule, a non-terminal can be replaced either by two non-terminals or by a terminal. When the prules in a context-free grammar have been so transformed, the grammar is said to be in *Chomsky normal form*. The language produced by the original context-free grammar and its Chomsky normal form is the same.

To transform a given grammar into Chomsky normal form, we transform the prules in the grammar one by one. If a prule in the grammar already satisfies the Chomsky normal form, we leave it as it is. Let there be a prule

$$I \longrightarrow \beta$$

in the grammar such that the prule is not in the Chomsky normal form. If β is a single non-terminal, delete this prule and, in all other prules, replace β by I. Otherwise, $|\beta| \geq 2$; hence proceed as follows:

1. If there is a terminal a in β, substitute a new non-terminal J for a and add a prule

$$J \longrightarrow a$$

Repeat this for each terminal in β: substitute a new non-terminal for each terminal and add a corresponding prule.

2. The string β will now have only non-terminals. If β contains two non-terminals, the prule is in Chomsky normal form. Otherwise, substitute a new non-terminal for each pair of non-terminals, and add a prule whose left-hand side contains the new non-terminal, and the right-hand side contains the two substituted non-terminals. Repeat this until all prules satisfy the Chomsky normal form.

To illustrate, let us transform the prule $I \longrightarrow KaLMb$ into Chomsky normal form. After substituting new non-terminals for the terminals and adding the required prules, we get the following prules:

1. $N \longrightarrow a$
2. $P \longrightarrow b$
3. $I \longrightarrow KNLMP$

Substituting new non-terminals for each pair of non-terminals in the right-hand side of the third prule, we get the following prules:

1. $N \longrightarrow a$
2. $P \longrightarrow b$
3. $Q \longrightarrow KN$
4. $R \longrightarrow LM$
5. $I \longrightarrow QRP$

The first four prules satisfy the Chomsky normal form but the fifth does not. Substituting a new non-terminal for a pair of non-terminals in the right-hand side of the fifth prule, we obtain the following prules:

1. $N \longrightarrow a$
2. $P \longrightarrow b$
3. $Q \longrightarrow KN$
4. $R \longrightarrow LM$
5. $T \longrightarrow QR$
5. $I \longrightarrow TP$

All the five prules are in Chomsky normal form.

11.3   Parsing

Given a grammar and a string $\delta = x_1 x_2 \ldots x_n$ of terminals, we are said to *parse* δ if we can determine whether δ is a sentence in the language produced by the grammar. The following Cocke–Younger–Kasami procedure (named after those who proposed it) can be used to parse δ when the grammar is context-free in Chomsky normal form. We are using S as the start symbol of the grammar. Assume that the prules in the grammar have been numbered 1, 2, 3, ...

1. Create a triangular table Z of columns $i = 1$ to n, and rows $j = 0$ to $(n + 1 - i)$, such as the one shown in Fig. 11.1. For $1 \leq i \leq n$ and $1 \leq j \leq (n + 1 - i)$, cell $Z(i, j)$ will be used to store non-terminals that can produce a string of length j starting from x_i.

2. For $i = 1, 2, \ldots, n$ do steps 2.1 and 2.2.

    2.1. $Z(i, 0) \leftarrow x_i$
          The n symbols of δ are thus stored in cells $Z(1, 0)$ to $Z(n, 0)$, one symbol in each cell.

    2.2. Put in $Z(i, 1)$, the non-terminals on the left-hand side of prules in the grammar that have the entry of $Z(i, 0)$ on their right-hand side. Associate with each non-terminal the number of the prule and the values i and 0 to indicate that the non-terminal is in a prule, which if fired, will produce the entry of $Z(i, 0)$.

3. For $j = 2, 3, \ldots, n$ do
    for $i = 1, 2, \ldots, (n + 1 - j)$ do
    for $k = 1, 2, \ldots, (j - 1)$ do
    put in $Z(i, j)$ the non-terminals on the left-hand side of prules that have on their right side an entry of $Z(i, k)$ concatenated on its right to an entry of $Z(i + k, j - k)$. Associate with each non-terminal the number of the prule and the values i, k, $(i + k)$, $(j - k)$ to indicate that the non-terminal is in a prule, which if fired, will produce an entry of $Z(i, k)$ concatenated on its right with an entry of $Z(i + k, j - k)$.

4. If start symbol S is in cell $Z(1, n)$, then conclude that the string $\delta = x_1 x_2 \ldots x_n$ is a sentence in the language produced by the given

grammar. The sequence of prules fired to produce δ can be obtained by the values associated with the non-terminals in Z.

Step 3 of the procedure describes how entries are made in cell $Z(i, j)$ for $2 \leq j \leq n$ and $1 \leq i \leq (n + 1 - j)$. In other words, enter those non-terminals in $Z(i, j)$ that exist on the left-hand side of prules which when fired will produce the concatenations of the entries of the cells in the following sequence:

$Z(i, 1)$ concatenated on its right with $Z(i + 1, j - 1)$
$Z(i, 2)$ concatenated on its right with $Z(i + 2, j - 2)$
$Z(i, 3)$ concatenated on its right with $Z(i + 3, j - 3)$
$\vdots$
$Z(i, j - 1)$ concatenated on its right with $Z(i + j - 1, 1)$

As an example, consider a context-free grammar whose terminals are a, b, and c; and non-terminals are S, A, B, and C, with S being the start symbol. The prules in Chomsky normal form are as follows.

1. $S \longrightarrow AC$
2. $S \longrightarrow BC$
3. $A \longrightarrow AC$
4. $A \longrightarrow b$
5. $B \longrightarrow AC$
6. $B \longrightarrow b$
7. $C \longrightarrow BB$
8. $C \longrightarrow a$

We have to parse string bab^2 to determine whether it can be produced by the grammar. Since there are four symbols in the string to be parsed, $n = 4$. The parsing is shown in Fig. 11.1. The caption of the figure explains how some of the entries are made in the triangular table.

Given a number of grammars and a string δ, we can use the Cocke–Younger–Kasami procedure to find out the language of which grammar δ belongs to. This is considered akin to finding the class to which a pattern belongs to, in the terminology used earlier in this book.

	$i=1$	$i=2$	$i=3$	$i=4$
$j=4$	$S,2,1,2,3,2$			
$j=3$	$C,7,1,2,3,1$			
$j=2$	$S,1,1,1,2,1$ $S,2,1,1,2,1$ $B,5,1,1,2,1$	$C,7,3,1,4,1$		
$j=1$	$A,4,1,0$ $B,6,1,0$	$C,8,2,0$	$A,4,3,0$ $B,6,3,0$	$A,4,3,0$ $B,6,4,0$
$j=0$	b	a	b	b

FIGURE 11.1 Parsing the string $babb$ for the grammar given in Section 11.3. We first fill the cells of the 0th row, then first row, then second row, and so on. Each cell in the 0th row of the above triangular table Z contains one of the terminals of $babb$. Terminal b in cell $Z(1,0)$ can be produced either by firing the fourth prule whose left-hand side is A, or by firing the sixth prule whose left-hand side is B. Thus, we store $A,4,1,0$ and $B,6,1,0$ in cell $Z(1,1)$. By storing $A,4,1,0$, we indicate that firing the fourth prule will produce b, the entry in cell $Z(1,0)$. We similarly fill the other cells in the first row. The entry $S,1,1,1,2,1$ in cell $Z(1,2)$ indicates that firing the first prule will produce an entry of cell $Z(1,1)$ that will be concatenated on its right by an entry of cell $Z(2,1)$. The string AC will be produced by A of $Z(1,1)$ concatenated on the right with C of $Z(2,1)$. Cell $Z(2,2)$ is empty because the prules do not have a non-terminal to produce concatenations of strings in $Z(2,1)$ and $Z(3,1)$. We next fill cell $Z(3,2)$. Cell $Z(1,3)$ should contain non-terminals that can produce strings from entries in $Z(1,1)$ concatenated on their right with the entries of $Z(2,2)$, and from entries in $Z(1,2)$ concatenated on their right with entries of $Z(3,1)$. Cell $Z(2,3)$ remains empty. $Z(1,4)$ should contain non-terminals that can produce strings concatenating entries of cells $Z(1,1)$ with $Z(2,3)$, cells $Z(1,2)$ with $Z(3,2)$, and cells $Z(1,3)$ with $Z(4,1)$. Since cell $Z(1,4)$ contains the start symbol S, string $babb$ can be produced by the given grammar. The numbers associated with each non-terminal in the table indicates the sequence of prules to be fired to produce the string from the grammar: $S \overset{2}{\Longrightarrow} BC \overset{5}{\Longrightarrow} ACC \overset{4}{\Longrightarrow} bCC \overset{8}{\Longrightarrow} bAC \overset{7}{\Longrightarrow} baBB \overset{6}{\Longrightarrow} babB \overset{6}{\Longrightarrow} babb.$

11.4   Stochastic Grammars

With each prule in grammar, if we associate a probability to indicate the probability of that prule firing, then we have a stochastic grammar. In the grammar used in the example given in Section 11.3, we have eight prules.

Let p_k be the probability of firing the kth prule. Let us assume that the firing of one prule is independent of the firing of any other prule. Then the probability of string *babb* being produced by firing the sequence of prules $S \overset{2}{\Longrightarrow} BC \overset{5}{\Longrightarrow} ACC \overset{4}{\Longrightarrow} bCC \overset{8}{\Longrightarrow} bAC \overset{7}{\Longrightarrow} baBB \overset{6}{\Longrightarrow} babB \overset{6}{\Longrightarrow} babb$ is

$$p_2 \times p_5 \times p_4 \times p_8 \times p_7 \times p_6 \times p_6$$

A probability is associated with each string in the language of a stochastic grammar. The probability indicates the probability of the string being produced by the grammar. This is akin to finding out the probability of a given pattern belonging to a given class, which was discussed in Sections 3.2 and 5.1.

Summary

There are four types of grammars: unrestricted, context-sensitive, context-free, and finite-state. In practice, context-free grammars are the ones most often used. The Cocke–Younger–Kasami procedure can be used to determine whether a particular string (sequence of symbols) has been produced by a given context-free grammar.

Exercises

1. In a programming language of your choice implement a procedure to transform a given context-free grammar into Chomsky normal form. Use this program to transform into Chomsky normal form the grammar with the following prules:

   (a) $S \longrightarrow aB$

   (b) $S \longrightarrow bA$

   (c) $A \longrightarrow b$

   (d) $A \longrightarrow cA$

   (c) $B \longrightarrow aA$

   In the above grammar, the terminals are a, b, and c; and the non-terminals are S, A, and B, with S being the start symbol.

2. Implement the Cocke–Younger–Kasami procedure in a programming language of your choice. Use the program to determine which of

the following strings can be produced by the Chomsky normal form grammar of the first exercise above:

  aacb

  bccac

  bccab

If the string can be produced by the grammar, the program should print the sequence of prules to be fired for producing the string; otherwise, it should print a message that the string cannot be produced by the given grammar.

Learning Objectives

This chapter contains the following topics:

- a comparison of the classifications of the professor–student recall patterns of the first chapter by different classification procedures
- the use of a committee of classifiers
- typical areas of application of pattern recognition techniques

12

Summing Up

The previous chapters discuss the various procedures used in pattern recognition. This chapter makes general comments about the procedures discussed in the earlier chapters.

Figure 12.1 recapitulates the classification results of the professor–student recall patterns given in Fig. 1.2, by classifiers adopting the procedures of SpecToGen, GenToSpec, decision tree built by maximizing the ratio of information gain, Naive Bayes, nearest neighbour, and neunet. The figure shows that the last four classifiers give identical results. This can tempt one to conjecture that perhaps the four classifiers, so seemingly different, have some underlying commonality[1] (detecting the commonality, if any, is a topic for research). Trained on a training set of another domain, the four classifiers may give different results for a given set of recall patterns.

Different classifiers giving different results should not be surprising. Humans often disagree with one another. In Section 1.3, when you asked your friends to intuitively classify the professor–student recall pattern, you must have received different classifications. Because different humans have different experiences over their life time, their training becomes dissimilar. Hence, given the same information, different humans may arrive at different conclusions. Two doctors looking at the same patient may diagnose differently and prescribe different treatment: after seeing one doctor, a patient sometimes seeks the opinion of another doctor (the remark that from two doctors you can get three opinions is a jocular exaggeration). At times, we may disagree with our colleagues on some issue. Although their conclusion may be different from ours, their reasoning may be as strong as ours. That is why, we can disagree, but we should not become disagreeable.

In a non-critical domain (differentiating professors from students, for instance), where occasional misclassification, is considered acceptable, any of the classifiers mentioned in this book can be chosen. In a critical domain, say diagnosing a life-threatening disease, one may want to employ many classifiers, and then choose the result confirmed by the majority. This is similar to having a committee of classifiers where decisions conform to majority vote. If we were to consider the six classifiers given in Fig. 12.1 as a committee, then by majority vote we would classify recall pattern R2 as professor, and the remaining three patterns as student. If we have more faith in some classifiers, we can give their votes more weightage than the votes

---

[1] The author is not aware of any proof that the four classifiers will always give identical results or that any classifier is more reliable than others.

Name of recall pattern	Attributes			Procedure used for classification		
						Decision tree, Naive Bayes, nearest
	HABIT	EATS	FOOT-WEAR	SpecTo-Gen	GenTo-Spec	neighbour, and neunet
R1	Quiet	Baked	Clogs	S	S	S
R2	Quiet	Roasted	Sandals	P	P	P
R3	Gabby	Roasted	Clogs	?	P/S	S
R4	Quiet	Roasted	Clogs	?	P/S	S

FIGURE 12.1 Recapitulating the classifications of the professor–student recall patterns given in Fig. 1.2 by different classifiers. P denotes a professor, S a student, and the question mark a rejection. The classifications by the SpecToGen classifier are reproduced from Fig. 1.4, and those by the GenToSpec classifier are from Fig. 1.6. SpecToGen rejects patterns R3 and R4. GenToSpec can classify each of R3 and R4 as professor or student, depending on which prule is selected to fire. The classifications by the decision tree are from Fig. 2.11 for the decision tree given in Fig. 2.9, the tree obtained by maximizing the ratio of information gain. The Naive Bayes classifications are from Fig. 5.2, the nearest neighbour classifications are from Fig. 6.1, and the neunet classifications are from Fig. 7.25. Coincidentally, the classifications by the decision tree, Naive Bayes, nearest neighbour, and neunet are identical, and hence they are shown in one column.

of those classifiers in which we have less faith—which is like a committee of humans where some people's opinions may carry more weightage than those of others. One may not like committees (according to one observation, a camel is a horse designed by a committee), but there are a lot of decisions taken by committees in this world.

If a committee of classifiers is used, then the programming of the different classifiers should be done by different persons or teams. It reduces the likelihood of any one person's or team's programming idiosyncrasies or misunderstandings infecting all the classifiers. If one classifier always confirms a result different from the other classifiers, then it may be desirable to take a closer look at the implementation of that classifier.

When human experts agree with the results output by a classifier for quite a few recall patterns, then consider adding the patterns to the existing training set and training the classifier again. The classifier will thus learn continuously, just as humans are expected to do. Over time, the classifier may start giving more and more reliable results. This, however, is a lengthy process. Humans, too, usually take a long time to become experts. Humans, however, die and all their expertise is lost. The classifiers, being software implementations, can be easily replicated, hence can go on and on.

It is apparent that in the nearest neighbour classifier, no computation is required in training, but a lot of computation is required in classifying a recall pattern since the distances that are required to be calculated are equal to the training patterns present. In contrast, other classifiers require considerable computation in training, be it building a decision tree or finding the appropriate weights for a neunet, but there is far less effort in classifying a recall pattern. In an informal sense, we face somewhat similar situations in our daily lives. As students, if we spend the desired effort in training ourselves for an examination, then we will usually find the examination easy and we may not have to spend much effort during the examination, the examination being analogous to a recall set. If, however, we do not spend the desired effort in training ourselves, then we will typically find the examination difficult, and we may have to spend considerable effort during the examination.

Humans are more versatile than classifiers. Nonetheless, fatigue, emotions, and boredom can sometimes dampen a human's objective performance. Classifiers do not face this restriction. Ultimately, the synergy of humans and computers will provide the world better and better classifiers in different domains. Some of the areas where classifiers find applications (review Section 1.6), are as follows.

- Optical character recognition of both printed and handwritten text, which can lead to the development of reading machines for the visually challenged.
- Signature verification, which may interest the banks.
- Speech recognition, by which a computer may serve as a stenographer.
- Weather prediction from satellite photographs.
- Fingerprint identification, which may interest the police.

- Geological exploration, in which, given the chemical composition of soil, the presence of minerals in that area may be predicted.
- Recognizing military installations, such as missile silos, from aerial photographs.
- Assessing credit risk from the background of an individual asking for a bank loan or credit card.
- Medical diagnosis, in which, given a patient's symptoms, the patients's disease is diagnosed and treatment prescribed.

Learning Objectives

This appendix provides a review of the following topics in probability theory:

- prior, conditional, and joint probabilities
- probabilities of different kinds of events: independent, mutually exclusive, completely correlated, and exhaustive
- binomial distribution
- Poisson distribution
- standard normal density function

A Review of Probabilities

While applying pattern recognition techniques, at times it is required to use probabilities. A review of probabilities is given below mainly to refresh the reader's memory but for readers who are new to probability theory, it is advisable to read a book on probability theory.

1. $P(Q)$ is the probability of event Q occurring. It is called the *prior*, or *a priori*, probability of Q. The occurrence of Q is not conditioned on the occurrence of any other event. If $P(Q) = 0$, then Q never occurs. If $P(Q) = 1$, then Q always occurs.

2. $P(Q_1|Q_2)$ is the probability of event Q_1 occurring conditioned on event Q_2 having already occurred. For brevity, it is often read as the 'probability of Q_1 given Q_2'. If Q_1 and Q_2 are independent events, that is, the occurrence of one event does not influence the occurrence of the other event, then

$$P(Q_1|Q_2) = P(Q_1)$$

3. $P(Q_1 \wedge Q_2)$ is the probability of events Q_1 and Q_2 occurring together. It is often read as the 'joint probability of Q_1 and Q_2'. By definition,

$$P(Q_1 \wedge Q_2) = P(Q_2 \wedge Q_1)$$

Moreover,

$$P(Q_1 \wedge Q_2) = P(Q_1|Q_2)P(Q_2) = P(Q_2|Q_1)P(Q_1)$$

Therefore,

$$P(Q_1|Q_2) = \frac{P(Q_1 \wedge Q_2)}{P(Q_2)} = \frac{P(Q_2|Q_1)P(Q_1)}{P(Q_2)}$$

If Q_1 and Q_2 are independent events, then

$$P(Q_1 \wedge Q_2) = P(Q_1)P(Q_2)$$

In general, if Q_1 to Q_n are independent events for $n \geq 1$, then

$$P(Q_1 \wedge Q_2 \wedge \ldots \wedge Q_n) = P(Q_1)P(Q_2)\ldots P(Q_n)$$

If Q_1 and Q_2 are mutually exclusive events, that is, both cannot occur at the same time, then

$$P(Q_1 \wedge Q_2) = 0$$

4. $P(Q_1 \vee Q_2)$ is the probability of events Q_1 or Q_2 or both occurring. By definition,

$$P(Q_1 \vee Q_2) = P(Q_2 \vee Q_1)$$

Moreover,

$$P(Q_1 \vee Q_2) = P(Q_1) + P(Q_2) - P(Q_1 \wedge Q_2)$$

If Q_1 and Q_2 are independent events, then $P(Q_1 \wedge Q_2) = P(Q_1)P(Q_2)$; therefore

$$P(Q_1 \vee Q_2) = P(Q_1) + P(Q_2) - P(Q_1)P(Q_2)$$

If Q_1 and Q_2 are mutually exclusive events, $P(Q_1 \wedge Q_2) = 0$, and therefore

$$P(Q_1 \vee Q_2) = P(Q_1) + P(Q_2)$$

In general, if Q_1 to Q_n are mutually exclusive events for $n \geq 1$, then

$$P(Q_1 \vee Q_2 \vee \ldots \vee Q_n) = P(Q_1) + P(Q_2) + \cdots + P(Q_n)$$

If Q_1 and Q_2 are completely correlated, that is, Q_1 occurs if, and only if, Q_2 occurs, then

$$P(Q_1 \vee Q_2) = P(Q_1) = P(Q_2)$$

If Q_1 to Q_n are exhaustive events for $n \geq 1$, that is, at least one of them must occur, then

$$P(Q_1 \vee Q_2 \vee \ldots \vee Q_n) = 1$$

If Q_1 to Q_n are exhaustive and mutually exclusive events for $n \geq 1$, then

$$P(Q_1 \vee Q_2 \vee \ldots \vee Q_n) = P(Q_1) + P(Q_2) + \cdots + P(Q_n) = 1$$

5. $P(\neg Q)$ is the prior probability of event Q not occurring. Since, Q and $\neg Q$ are exhaustive and mutually exhaustive, $P(\neg Q) + P(Q) = 1$, therefore

$$P(\neg Q) = 1 - P(Q)$$

6. The *odds* of an event Q is defined to be

$$O(Q) = \frac{P(Q)}{1 - P(Q)}$$

By algebraic manipulation of the above equation

$$P(Q) = \frac{O(Q)}{1 + O(Q)}$$

7. Suppose an experiment has two possible outcomes: success or failure. Let the probability of success be θ. If the experiment is conducted $n \geq 1$ times, then the probability of $0 \leq k \leq n$ successes is given by the *binomial distribution*

$$\frac{n!}{k!(n-k)!}\theta^k(1-\theta)^{n-k}$$

8. Suppose an event occurs μ times on an average in some given time interval; for example, six cars on an average pass a given point on the road in a minute. Then the probability that the event will occur k times in that time interval is given by the *Poisson distribution*

$$\frac{e^{-\mu}\mu^k}{k!}$$

Using the preceding equation, we can, in our example, estimate the probability of the eight cars passing that point in the road in a minute: for this, substitute 6 for μ and 8 for k in the equation.

9. Suppose a continuous variable u represents an event, such as the height of a person. Then the probability distribution is described by a probability density function $p(u)$, such that

$$\int_{-\infty}^{\infty} p(u) = 1$$

Then the probability of u within some interval $[a, b]$ is

$$P(a \leq u \leq b) = \int_{a}^{b} p(u)$$

A commonly used density function is the *normal* (or *Gaussian*), *density* function

$$p(u) = \frac{1}{\sigma\sqrt{2\pi}}e^{-(u-\mu)^2/2\sigma^2}$$

where μ is the mean, σ is the standard deviation, and σ^2 is the variance of x. When $\mu = 0$ and $\sigma = 1$, the function is called *standard normal density*:

$$p(u) = \frac{1}{\sqrt{2\pi}}e^{-u^2/2}$$

Learning Objectives

This appendix proposes designing a classifier for breast cancer prognostication. To do that, it discusses the following topics:

- preliminary information about breast cancer and factors influencing it
- kinds of therapy for breast cancer patients: surgery, radiation, chemo, and hormone
- side effects of therapy
- attributes to be used for breast cancer patients
- requirements of the breast cancer prognosticator

B

Project on Breast Cancer Prognostication

B.1 Intent

To apply the classifiers described in this book to any domain, it is important to understand the domain well. In this appendix, as an example, a medical domain is given, for which a classifier has to be designed. While designing the classifier, give justifications for the design choices made.

Not all doctors will agree with what is written below, but suppose the following information has become available[1].

B.2 Breast Cancer: Preliminaries

In the human body, it sometimes happens that a few cells (units of organisms) begin to grow out of control to form a tumour. If the tumour stays in its region, it is usually benign. If, however, the tumour spreads to the distant parts of the body, then it is often malignant, or cancerous. The tumour is said to have *metastasized*, and the secondary tumour is called *metastasis* (plural, *metastases*). The disease is transferred from one part of the body to another through the lymph (an almost colourless fluid containing white blood cells) and the blood vessels.

One kind of cancer, namely breast cancer (BC), whose estimated frequency of occurrence is shown in Fig. B.1, is among the leading causes of death in women belonging to the age group of 35 to 54 years. BC usually spreads from the breast to the lymph nodes (bean-shaped structures) in the armpits. As men get older and their chest muscles sag, they, too, develop 'breasts'. BC in men is rare, but not unknown. For this project, however, let us restrict ourselves to women patients.

- *Age* BC risk increases with age. Women over 50 years of age constitute 75% of BC cases. Tumours in younger, pre-menopausal women, however, grow faster.
- *Family History* A woman whose mother or a sister has BC is considered to have a family history of the disease. BC patients with such family history are typically fifteen years younger than patients without such family history.

---

[1]This information was collected by the author with the help of his former students Micheline Kamber and Jennifer Scott.

By age	Estimated number of women out of which one woman is likely to develop breast cancer
30	622
45	93
55	33
65	17
75	11
85	9
95	8

FIGURE B.1   Estimated frequency of occurrence of breast cancer.

- *Menstrual History*   A girl beginning to menstruate before the age of 12 is one-and-a-half times more likely to develop BC than a girl who begins to menstruate after the age of 15. A woman who has menopause after the age of 55 is twice more likely to develop BC than a woman who has menopause before the age of 45. Overall, a woman who menstruates for a longer time faces a higher risk of BC. This is thought to be due to the woman's increased exposure to oestrogen, a hormone. The endocrine glands (thyroid, pituitary, adrenal, ovaries) produce chemicals called hormones, which are secreted directly into the blood.

- *Child Bearing History*   A woman who has her first pregnancy early, especially before the age of 25, is less likely to develop BC than a woman otherwise. If her first pregnancy is after the age of 30, then the woman increases the likelihood of developing BC by 17% to 20%. Such a woman is more likely to develop BC than a woman who has never been pregnant. A woman who has her first pregnancy after the age of 35 is two-to-three times more likely to develop BC than a woman who has her first pregnancy before the age of 20.

- *Breast Feeding*   If a woman has her first child before the age of 30, and she breast feeds her baby for at least six months, then the likelihood of her developing BC is reduced. Most BCs start in milk ducts. About 15% start in lobules at the end of the ducts.

- *Previous Breast Cancer*   If a woman has cancer in one breast then the likelihood of her developing cancer in the other breast increases by 25%. Moreover, she is five-to-six times more likely to develop another cancer.
- *Fibrocystic Breast Disease*   A woman with cysts in her breasts is more likely to develop BC in later life.
- *HER2/neu*   High levels of HER2/neu—a proto-oncogene that produces a protein that, in turn, stimulates growth of cells—indicate that the tumour will grow quickly.
- *Cathepsin D*   A woman with a rapidly growing tumour and high levels of Cathespin D—a protein secreted by tumour cells—has a 60% probability of cancer recurrence.
- *Synthetic Oestrogen Intake*   A post-menopausal woman taking oestrogen as hormone replacement treatment reduces the likelihood of heart disease and osteoporosis, but increases by 10% to 30% the likelihood of her developing BC.
- *Radiation Exposure*   A woman whose neck, chest, or breasts have been exposed to any kind of radiation increases her likelihood of developing BC.
- *Environmental Causes*   A woman exposed to toxic chemicals, pesticides, contaminated water, or industrial carcinogens increases her likelihood of developing BC.
- *Diet*   A diet high in fat is suspected to increase the likelihood of developing BC. Exercises will reduce fat, which may decrease the likelihood.
- *Check-ups*   A woman should examine herself regularly for tumours in her breasts, and she should have periodic mammograms, which are X-ray images of the soft tissues of the breasts. Early detection of BC reduces the likelihood of death by it. The dense breast tissue in younger women can make tumours hard to detect.
- *Benefits and Side Effects of Therapy*   BC recurs in 30% of women after therapy. The following percentage of women survive five years after diagnosis: 94% for those with early-detected, localized BC; 73% for those whose cancer has spread to the lymph nodes; 18% for those whose cancer has metastasized to distant sites. The death rate for women under the age of 50 is fifteen times that of women over the age of 50, BC

in younger women often being more aggressive. Whereas therapy for BC may prolong a woman's life, it may also have side effects that deteriorate her quality of life. In the long term, the therapy can have side effects even after the therapy has stopped, for example, damage to the heart, lungs, and kidneys. Therapy can even cause sterility and further cancer. Optimal therapy, which cures the patient without possibly harming her, is still unknown.

B.3   Therapy: Surgery and Radiation

When a woman has been diagnosed as having BC, the preparatory step towards treatment consists of establishing the extent of the cancer. Depending on the extent of the cancer, the patient usually has to undergo surgery and radiation therapy.

The simplest surgery is a *lumpectomy*, which removes the tumour with a margin of healthy tissues. The most extensive surgery is *radical mastectomy*, which removes the breast, muscles of the chest wall below the breast, and tissues in the armpits. This surgery is becoming infrequent nowadays. Between these two extremes lie other kinds of surgery: *partial* (or *segmental*) *mastectomy* removes a part of the breast; *simple* (or *total*) *mastectomy* removes the full breast and sometimes a few armpit tissues; and *modified radical mastectomy* removes the breast, armpit lymph nodes, and the lining over the chest muscles. Also known as *mastectomy with axillary* (or *underarm*) *dissection*, the last kind of surgery is reportedly the most common nowadays, especially for pregnant women. Removal of a breast, however, causes some women to feel that they have lost a part of their femininity.

Surgery is often followed by radiation therapy, in which radiation from radioactive substances is used to destroy cancerous cells. The patient may also be prescribed radio-sensitizers, which are drugs to boost the radiation therapy. Radiation therapy can also destroy healthy cells near the cancerous cells. The temporary side effects are fatigue, skin rash, and sensitivity in the treated areas.

B.4   Therapy: Chemo and Hormone

Besides surgery and radiation, drugs may be prescribed to work on the entire body. This is called *adjuvant* (or *systemic*) *therapy*. Typically started four weeks after surgery, it is given to those with cancerous lymph nodes, because that suggests cancer may be circulating in the body. There are three kinds of adjuvant therapy: *hormone* (or *endocrine*) *therapy, chemotherapy,* and *chemo-endocrine* therapy, the last being a combination of the first two.

In adjuvant hormone therapy, the objective is to deprive cancer cells of the hormones needed for growth. Hormones stimulate cells and blood tissues. The two hormones of interest are oestrogen and progesterone. After surgical removal, the breast tumours are tested for *hormone receptors.* Tumours that need the hormones to grow are said to be *receptor-positive,* otherwise *receptor-negative.* Tamoxifen, an anti-oestrogen drug, is given orally for two to five years to block the effects of the patient's own hormones on the cancer cells. Tamoxifen reduces recurrence by 40%. About two-thirds of women have oestrogen-receptor-positive tumours, and they are more likely to benefit from hormone therapy than those with oestrogen-receptor-negative tumours. A second test is for progesterone receptors. A patient with a tumour positive for both oestrogen and progesterone receptors has an 80% likelihood of benefiting from hormone therapy. Tamoxifen is usually tolerated well, but the lower oestrogen levels can cause hot flashes. Taking Tamoxifen for more than 5 years shows no further benefit, but instead can cause endometrial (lining of the uterus) cancer. After five years of Tamoxifen, a patient may be prescribed a drug called Letrozole for life. Letrozole reduces the likelihood of the cancer recurring by half.

In adjuvant chemotherapy, three to six cytotoxic drugs (also known as *cytolytic chemicals,* ones that destroy cells) are given orally or intravenously to kill the cancer cells. The therapy may last six to twelve months. Chemotherapy is intended to destroy cancer cells throughout the body. In general, chemotherapy reduces by about one-third the likelihood of death due to BC. Prednisone, a steroid, is also sometimes given to enhance the effects of the cytotoxic drugs. Chemotherapy is usually given in following cases:

- Tumour larger than 10 mm
- Four or more cancerous lymph nodes
- Hormone-receptor-negative tumours
- Fast-growing tumours
- High levels of cathepsin D or HER2/neu
- Age less than forty

At times, chemotherapy is done before surgery to reduce the size of the tumour, thus requiring less surgery. This is known as *neo-adjuvant chemotherapy*. In destroying cancer cells, chemotherapy may destroy healthy cells as well. The side effects of chemotherapy are the following:

- Temporary nausea and vomiting
- Ulcers in the mouth
- Loss of appetite
- Temporary loss of hair
- Anaemia (reduced red blood cells) causing fatigue, weakness, chills, dizziness, and shortness of breath
- Greater propensity to infections
- Reduction in platelets, which are responsible for clotting, causing bruising and bleeding
- Neuropathy resulting in double or blurred vision, headaches, drooping eyelids, dizziness, and numbness or tingling in the fingers and toes
- Allergies causing acne, redness and itching, dry skin, and brittle nails
- Weight gain
- Early menopause, especially for those near the age of 50
- Temporary infertility

B.5   Building a Training Set

Suppose there exists a cancer clinic that has maintained a record for each patient who has undergone therapy at that clinic. There are 22 items in the record of each patient. The first twenty items, which give details about the patient and the therapy she received, constitute the attributes of the patient. The next two items classify the patient according to the success of her therapy and the side effects she suffered.

The 22 items are described below. The first part of the project asks us to imagine that we work for the clinic. We need to decide how to code the values of these items in the record of a patient. The set of records for the different patients would constitute a file containing the training set to be used in the second part of the project.

1. What is the age of the patient, rounded off to the nearest year?

2. Did her mother or sister have BC?

3. At what age did she begin menstruating?

4. Is she pre-menopausal or post-menopausal? If she is post-menopausal, at what age did she have her menopause? This attribute value needs to be appropriately coded. For instance, a 0 could indicate that the patient is pre-menopausal. A non-zero value, say 49, would indicate her age at menopause. There are other attributes below, which need to be coded in a similar manner. The coding is left to the reader as an exercise. In general, assign a unique code for every possible value of an attribute.

5. Has she been pregnant? If yes, what was her age at first pregnancy? If she has had a child, did she breast feed the child for at least six months?

6. If she has cancer in one breast, did she have it earlier in the other breast?

7. Has she ever taken synthetic oestrogen supplements?

8. Has she been exposed to radiation?

9. Has she been exposed to toxic chemicals, pesticides, contaminated water, or industrial carcinogens?

10. How long ago did she have her last mammogram when no tumour was detected? This indicates for how long at the most the tumour has been growing. While coding this attribute value, allow for the situation where the patient has never had a mammogram.

11. What is the size of her tumour in millimetres?

12. How many cancerous lymph nodes does she have?

13. Is her oestrogen-receptor status positive or negative?

14. Is her progesterone-receptor status positive or negative?

15. Is her level of HER2/neu high?

16. Is her level of Cathepsin D high?

17. For how many months was she given neo-adjuvant chemotherapy?

18. What kind of surgery was done on her: none, lumpectomy, partial mastectomy, simple mastectomy, modified radical mastectomy, or radical mastectomy?

19. For how many months was she given adjuvant hormone therapy?

20. For how many months was she given adjuvant chemotherapy?

21. After surgery, for how many years (rounded off to the nearest year) was the patient free of cancer? If the patient remained free of cancer for 15 years, she is considered to have been fully cured. A value of more than 15 years is changed to 15 years, the integer value thus varying from 0 to 15. This means there will be 16 classes—let us call them *curative classes*—which can be named, say, curative-class 0 to curative-class 15.

22. As side effects of the therapy, which of these (there can be more than one) did the patient have: none, feeling of a loss of her femininity, hot flashes, nausea, vomiting, ulcers in the mouth, loss of appetite, loss of hair, anaemia, fatigue, weakness, chills, dizziness, shortness of breath, reduction in platelets, double or blurred vision, headaches, drooping eyelids, dizziness, numbness or tingling in fingers or toes, acne, redness and itching, dry skin, brittle nails, weight gain, infertility, cessation of menstruation. The last two will not apply to a patient who has already had her menopause. We will need to define classes—let us call them *offshoot classes*—according to the various combinations of the side effects a patient can have.

The first ten attributes are about the history of the patient, which she will provide to the clinic. Attributes eleven to sixteen are the results of the medical tests conducted on her by the clinic. Attributes seventeen to twenty report the therapy she received. The twenty-first and twenty-second items classify the patient according to the success of the therapy and the side effects she suffered. Every patient will fall into one of the curative classes and into one of the offshoot classes.

B.6   Designing a Breast Cancer Prognosticator

Let us suppose that a training set exists according to the description in Section B.5 and according to how the values of the items described in that section are coded. Design a classifier (any classifier described in this book can be chosen) that can be trained on that training set.

It is possible that, for a given patient, the values of some of the attributes are not known: for instance, a patient may not know whether she has ever been exposed to radiation (eighth attribute), or, say, owing to some oversight, the patient's progesterone-receptor status (fourteenth attribute) may not have been entered in her record. This is a case of missing attribute values, which is discussed in Chapter 3.

Once the classifier is trained, it can be used as a BrEast CAncer Prognosticator (BECAP). Given the values of the first 16 attributes of a new patient (a recall pattern, according to the terminology used in the book) and her proposed therapy (attributes 17 to 20), BECAP should be able to predict the number of years she is expected to be free of cancer (that is, assign the woman to one of the curative classes), and the side effect she is expected to suffer (assign her to one of the offshoot classes you set up). It is as if BECAP fired the following prule:

> If the age of the patient is ... ,
> and her family history of BC is ... ,
> ⋮
> and her last tumour-free mammogram was ... months ago,
> and the size of her tumour is ... millimetres,
> and she has ... cancerous lymph nodes,
> ⋮
> and her level of Cathepsin D is ... ,
> and she has been given neo-adjuvant chemotherapy for ... months,
> and the surgery done on her is ... ,
> ⋮
> and she is given adjuvant chemotherapy for ... months,
> then she is expected to be free of cancer for ... years,
> and she is expected to suffer these side effects: ... .

BECAP can thus be used to examine alternative therapies and then prescribe the best possible therapy for a given patient.

Learning Objectives

This appendix proposes implementing an optical character recognizer. To do that, it discusses the following topics:

- the character patterns required for the project
- implementing a recognizer for the patterns given in the Online Resource Centre of the book
- removing noise from the patterns implementing an improved recognizer
- skeletonizing the patterns and implementing a further improved recognizer
- using the frequency of letter occurrence to recognize text constructed from the patterns
- using the frequency of occurrence of character-pairs to improve the recognition of text

C

Project on Optical Character Recognition

@ C.1   Online Resource Centre

The Online Resource Centre of this book contains two ASCII files named train.txt and recall.txt. The files contain patterns of handwritten characters (capital letters A to Z, and digits 0 to 9) written by different people. We can say that the files contain patterns of classes C_1 to C_{36}. Figure C.1 shows a specimen pattern: a label above the pattern identifies the pattern to be of the letter A.

Each pattern is an image containing black and white pixels. The black pixels are represented by asterisks, and the white by blanks. The readers can copy the files into their computer and scroll through the two files to see what the other patterns look like. Each file contains more than one pattern of

```
A            ******
           ********
          **********
         *****   ****
         *****   ****
         ***     ****
         ***     ****
         ***      ***
         ***     ****
         ***    *****
         **************
         **************
         **************
         ******   ****
         ***      ****
         ***     ***
         ***     ****
         ******   ****
         *****    ****
         *****   ****
         ****
         **
```

FIGURE C.1   A specimen pattern of a handwritten letter. The patterns contains black and white pixels, asterisks representing the black pixels, and blanks the white pixels. The label identifying the pattern is at the top left.

each of the thirty-six characters. Patterns of the same character may not be identical to one another, because different people write differently. Moreover, patterns of different characters may look similar because of the way some people write; for example, B and 8; D, O (oh), and 0 (zero); I (the letter I written without serifs) and 1; S and 5; Z and 2.

Think of the attributes that help in distinguishing one handwritten character from another. The letter H ideally consists of two vertical strokes joined in the middle by a horizontal stroke, but because of the way a person wrote, the strokes may not be perfectly vertical, but only somewhat so. Accordingly, if a vertical stroke is used as an attribute, then some minor deviation would have to be accepted from the way in which a stroke should ideally be written. In general, minor deviations from the ideal would have to be accepted for any other attributes chosen. We need to consider not only the presence of a given attribute, but also where it is placed in the pattern. In the letter H, one vertical stroke should be on the left, one on the right, and the horizontal stroke should be in the middle.

C.2 Recognition Without Preprocessing

1. Select some attributes. The ratio of information gain provided by a given attribute (Section 2.5) can be used to decide whether to select a particular attribute.

2. In a language of your choice, write a program or a sequence of programs to do the following.

    2.1. Read the patterns of the file train.txt and, from each pattern, extract the attributes selected in step 1.

    2.2. Using the attributes extracted, train any classifier (decision tree, evolutionary, Bayes, nearest neighbour, neural net, linear) on the patterns of the file train.txt.

    2.3. Use the classifier to classify the patterns of the file train.txt. If the classification of a pattern matches the label of the pattern, then the pattern has been recognized correctly, otherwise incorrectly. Display the recognition rate (the proportion of patterns classified correctly) and the confusion matrix. In a confusion matrix con, the element $con(i, j)$ indicates the number of patterns of class C_i that were classified into class C_j, that is, it shows the confusion between classes C_i and C_j. If classes C_i and C_j are often confused, then consider using attributes that help in distinguishing between the two classes. A classifier with a 100% recognition rate will have its confusion matrix to be a diagonal matrix. The confusion matrix will have 36 rows and 36 columns. If the recognition rate or the confusion matrix is not satisfactory, then go to step 1 to select different attributes, or go to step 2.2 to develop a different classifier. For instance, if a neural net is being used, consider each pixel to be an attribute having two possible values, black or white.

3. Read the patterns of the file recall.txt and, from each pattern, extract the attributes selected in step 1. Use the classifier trained in step 2.2 to classify the patterns of the file recall.txt. Display the recognition rate and the confusion matrix.

C.3   Removing Noise

To obtain a pattern, a character written on a sheet of paper is scanned using a camera. Ideally, the black points in the character should become black pixels in the pattern, and the white points the white pixels. This, however, does not always happen in practice. Because of faintness or breaks in character strokes, smudges in the vicinity of the character, dirt on the lens of the camera, and flaws in the camera lighting, the following types of *noise* are introduced in the pattern:

- *Salt* noise: These are pixels that are white in the pattern, but they should have been black.
- *Pepper* noise: These are pixels that are black in the pattern, but they should have been white.

Noise can worsen a pattern's quality, which in turn can increase the difficulty of recognizing it correctly. It is often recommended that noise be removed as far as possible before attributes are extracted from the pattern. Noise removal is considered to be part of *preprocessing* a pattern.

To decide whether a given pixel p is noise, we need to look at the 3 by 3 window around it. The window contains the pixel p and the eight pixels q_0

q_3	q_2	q_1
q_4	p	q_0
q_5	q_6	q_7

FIGURE C.2   A pixel p and its neighbourhood of adjacent pixels $q_0, q_1, \ldots , q_7$ in a 3 by 3 window.

to q_7 adjacent to it (Fig. C.2). Pixels q_0 to q_7 constitute the *neighbourhood* of pixel p.

Let us consider a pixel to have a Boolean value TRUE if it is black, and FALSE otherwise. To remove salt and pepper noise, we make a single pass over the pattern scanning row by row, pixel by pixel. In other words, each pixel is considered to be p by turn. To remove some of the salt noise[1], a white pixel p is *filled* (changed to black) if the following Boolean expression is true:

$$[q_0 \wedge q_1 \wedge q_2 \wedge (q_4 \vee q_5 \vee q_6)]$$
$$\vee \, [q_2 \wedge q_3 \wedge q_4 \wedge (q_6 \vee q_7 \vee q_0)]$$
$$\vee \, [q_4 \wedge q_5 \wedge q_6 \wedge (q_0 \vee q_1 \vee q_2)]$$
$$\vee \, [q_6 \wedge q_7 \wedge q_0 \wedge (q_2 \vee q_3 \vee q_4)]$$

As is usual in Boolean expressions $\wedge$ denotes logical AND, and $\vee$ denotes logical OR. We can informally say that, in a white pixel p's neighbourhood, if one corner has three black pixels, and its opposite corner has at least one black pixel, then fill p.

To remove some of the pepper noise, a black pixel p is *deleted* (changed to white) if the following Boolean expression is false:

$$[(q_0 \vee q_1 \vee q_2) \wedge (q_4 \vee q_5 \vee q_6)]$$
$$\vee \, [(q_2 \vee q_3 \vee q_4) \wedge (q_6 \vee q_7 \vee q_0)]$$

---

[1]The procedure to remove salt noise is given in Iliescu, S., Shinghal, R., and Teo, R. Y. 'Proposed heuristic procedures to preprocess character patterns using line adjacency graphs', *Pattern Recognition*, vol. 29, no. 6, 1996, pp. 951–75.

This informally means that, in two non-opposite corners of a black pixel p's neighbourhood, out of the three pixels in each corner, if no pixel is black, then delete p. This method of removing pepper noise is known as 'Unger's method' after S.H. Unger who had proposed it.

While considering a pixel p in the row or column bounding the pattern, we will need to imagine a white row or column adjacent to p; for example, if p is in the topmost row of the pattern, imagine that pixels q_1, q_2, and q_3 are white.

In any chosen language, write a program to remove the salt and pepper noise from the patterns of the files train.txt and recall.txt. Save the patterns after removing the noise in corresponding files clean_train.txt and clean_recall.txt. Repeat the exercise done in Section C.2 except now train on the file clean_train.txt and test on clean_recall.txt. Compare the classification results with those obtained in Section C.2.

C.4 Skeletonization

Think of what would happen if someone stopped eating for quite a few days. In due course, the person will be thinned to a skeleton.

Skeletonizing[2] a pattern consists of deleting the black pixels along the edges of the strokes of the pattern until the pattern is thinned to a line drawing, which is called the pattern's *skeleton*. Figure C.3 shows the pixels to be deleted from the pattern shown in Fig. C.1 to obtain the skeleton shown in Fig. C.4. The skeleton retains the essential shape of the pattern. Whereas the strokes in a pattern can have varying thickness, the strokes in its skeleton should ideally have a uniform thickness of one pixel.

It has been observed that, in practice, we can often get better classification results if, as part of preprocessing, we first remove noise from the patterns, and then skeletonize them. We then train the classifier on the skeletons of the training patterns. This means we extract attributes from the skeletons, not from their patterns. Since the skeletons have greater uniformity than their patterns, it is expected that the attributes extracted from the skeletons will

---

[2]Skeletonization by SPTA is given in Naccache, N.J. and Shinghal, R., 'SPTA: A proposed algorithm for thinning binary patterns', *IEEE Transactions on System, Man, and Cybernectics*, vol. SMC-17, no. 3, 1984, pp. 409–18.

```
A          . . . . . .
         . . * * * * . .
       . . * . . . . * . .
     . . * . . . . . . * . .
   . . * . . . . . . . * . .
   . * . . . . . . * . .
   . * . . . . . . * . .
   . * . . . . . . * . .
   . * . . . . . . * . .
   . * . . . . . . * . .
   . * . . . . . . . * . .
   . * . . . . . . . . * . .
   . * * * * * * * * * . . .
   . * . . . . . . . * . .
   . * . . . . . * . .
   . * . . . . . * . .
   . * . . . . . * . .
   . . * . . . . * . .
   . . * . . . . * . .
   . . * . . . . * . .
   . . * . . . . * . .
   . . * . . . . * . .
   . . * . .     . * . .
   . . * .
   . *
```

FIGURE C.3   Deleting pixels shown as points from the pattern given in Fig. C.1 will produce a skeleton of the pattern. The asterisks will constitute the skeleton, which is shown in Fig. C.4.

have greater uniformity as well, hence the training will improve. To classify a recall pattern, we first preprocess it (remove its noise and skeletonize it), extract attributes from it, and then classify it.

The decision on selecting the pixels to be deleted from the pattern shown in Fig. C.1 was taken after a visual inspection of the pattern. In general, however, we need a formal criterion[3] for selecting the black pixels to be deleted. The criterion can then be incorporated in a skeletonization algorithm. Quite a few skeletonization algorithms exist. In this book, however, only one of them called SPTA (Safe Pixel Thinning Algorithm)

[3]A refinement to SPTA's terminating criterion appeared in Beffert, H. and Shinghal, R., 'Skeletonizing binary patterns on the homogeneous multiprocessor', *International Journal of Pattern Recognition and Artificial Intelligence*, vol. 3, no. 2, 1989, pp. 207–16.

```
A          ****
         *     *
        *       *
       *        *
      *         *
     *          *
     *          *
     *          *
     *          *
     *          *
     ********** *
     *          *
      *         *
      *         *
      *         *
      *         *
      *         *
       *        *
       *        *
       *        *
        *
        *
```

FIGURE C.4   A skeleton of the pattern shown in Fig. C.1. The skeleton was obtained by deleting pixels along the edges of the pattern as shown in Fig. C.3.

is described. To begin with, the definitions of terms used in SPTA are presented, then the underlying principles of SPTA, and then how SPTA works.

The eight pixels q_0 to q_7 adjacent to a pixel p, as shown in Fig. C.2, are termed to be the *neighbours* of p (do not confuse the meaning of the term *neighbour* here with its meaning in the nearest neighbour classifier discussed in Chapter 6). Within these eight neighbours of p, pixels q_0, q_2, q_4, and q_6 are more specifically said to be the *orthogonal* neighbours of pixel p.

A pattern is *connected* if, between every pair of black pixels c_0 and c_n, we can find a sequence of black pixels c_0, c_1, c_2, ... , c_n such that c_{i-1} is a neighbour of c_i, for $1 \leq i \leq n$. A *break pixel* is a black pixel whose deletion would cause a connected pattern to no longer remain so.

An *end pixel* is a black pixel at the end of a stroke. SPTA formally defines it to be a black pixel that has at most one black neighbour.

Edge pixels are black pixels lying along the edges of the strokes of a pattern. SPTA formally defines an *edge pixel* to be a black pixel that has at least one white orthogonal neighbour. There are four kinds of edge pixels:

A *right-edge pixel* has its right orthogonal neighbour q_0 white.
A *top-edge pixel* has its top orthogonal neighbour q_2 white.
A *left-edge pixel* has its left orthogonal neighbour q_4 white.
A *bottom-edge pixel* has its bottom orthogonal neighbour q_6 white.

An edge pixel can be of more than one kind; for example, a black pixel that has neighbours q_0 and q_2 white is both a right-edge pixel and a top-edge pixel.

In essence, SPTA executes passes over a given pattern and in each pass flags edge pixels that

- are not end pixels, for deleting these will delete the ends of strokes;
- are not break pixels, for deleting these will break the connectedness of the pattern; and
- do not cause *excessive erosion* by iteratively deleting a stroke.

At the end of each pass, SPTA deletes the flagged pixels. SPTA stops when there are no more pixels to be flagged. The black pixels remaining constitute the skeleton.

Consider windows (a) to (d) shown in Fig. C.5, in which the x's and y's are *immaterial pixels:* it does not matter whether they are black or white. As discussed below, if the neighbourhood of a left-edge pixel p matches any of the windows (a) to (d), then p is not flagged for deletion during skeletonization.

1. If the neighbourhood of p matches any of the windows (a) to (c), then

    1.1. If all x's are white, then p becomes an end pixel; therefore p should not be flagged.

    1.2. If at least one x is black, then p becomes a break pixel; hence p should not be flagged.

2. If p matches window (d), then

    2.1. If at least one x and one y are black, then p becomes a break pixel; hence p should not be flagged.

2.2. If all x's are white, then the eight possible neighbourhoods of p are shown in windows (d1) to (d8). Then

    2.2.1. Windows (d1) to (d3) make p an end pixel; hence p should not be flagged.

    2.2.2. Window (d4) makes p a break pixel; hence p should not be flagged.

    2.2.3. Pixel p should not be flagged for deletion in windows (d5) and (d6) because it can be empirically shown that deleting p will cause excessive erosion: a slanting stroke of thickness 2, such as (e), will get iteratively erased to become (f). By not deleting p, stroke (e) will iteratively become (g), which is the shape the skeleton of (e) should have.

    2.2.4. A black pixel p in windows (d7) and (d8) is pepper noise, and p should have been deleted by Unger's pepper-noise removal procedure described in Section C.3. It is mentioned earlier that noise should be removed before skeletonization. Hence, windows (d7) and (d8) will not exist at the beginning of skeletonization. If window (d7) occurs in an intermediate stage of skeletonization, then p would be a bulge in the pattern: to retain the shape of the pattern, p should not be flagged. If window (d8) occurs in an intermediate stage of skeletonization, then p would be an isolated pixel: p should not be flagged to retain the shape of the pattern. Window (d8) can occur in an intermediate stage of skeletonizing the pattern of lower case i, for instance, because of the separate dot at the top.

2.3. If all y's are white, then the eight possible neighbourhoods of p are symmetric to windows (d1) to (d8), and it can be argued, as in items 2.2.1 to 2.2.4, that p should not be flagged.

If a left-edge pixel p does not match any of the windows (a) to (d) shown in Fig. C.5, it is flagged for deletion. This is equivalent to saying p is flagged for deletion if the Boolean expression E_L is true, where

$$E_L = q_0 \wedge (q_1 \vee q_2 \vee q_6 \vee q_7) \wedge (q_2 \vee \neg q_3) \wedge (q_6 \vee \neg q_5)$$

A pixel is true if it is black and unflagged; otherwise it is false (that is, the pixel is false if it is white, or if it is black and flagged). This definition of the Boolean value of a pixel differs slightly from that used in removing noise in Section C.3. If E_L is false for a left-edge pixel p, then p is declared to be a *safe pixel*; it is not flagged for deletion.

Boolean expressions can be similarly developed for other kinds of edge pixels.

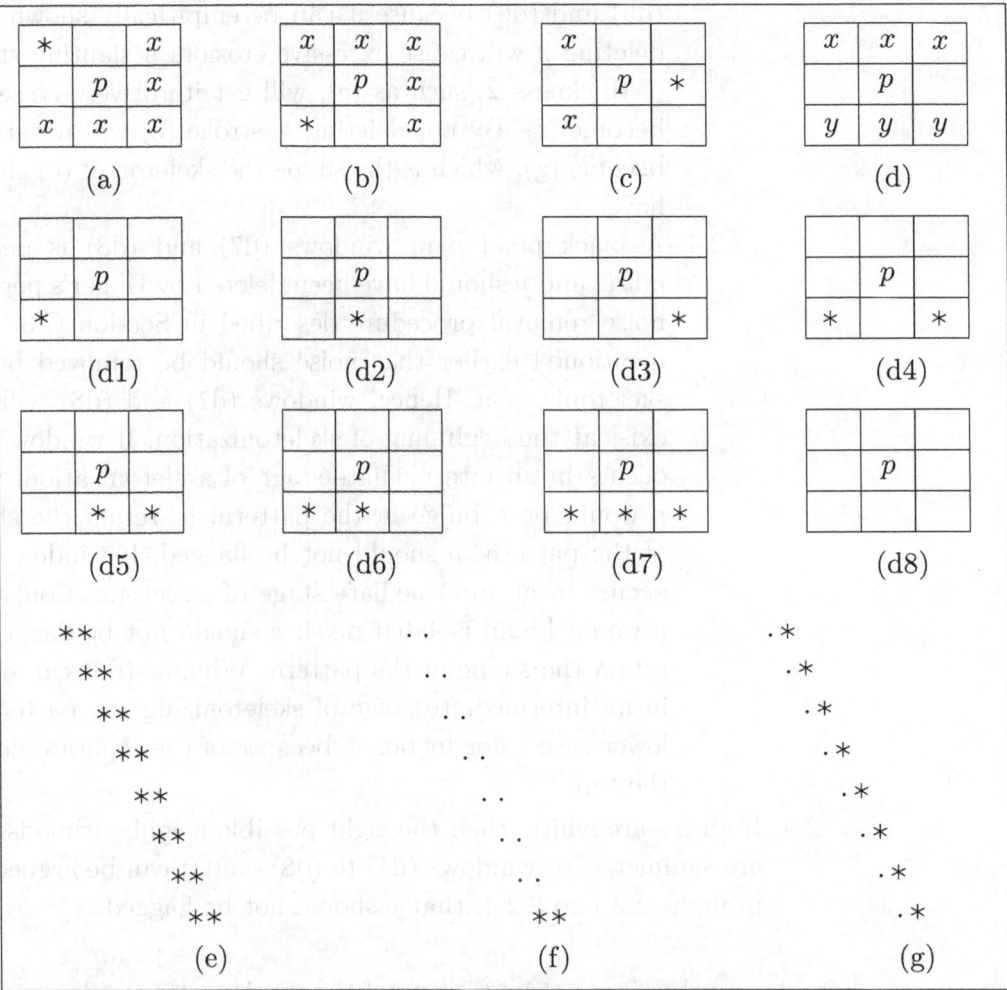

FIGURE C.5   During its pass over a pattern, SPTA does not flag a left-edge pixel p if its neighbourhood matches any of the four windows (a) to (d). The x's and y's are *immaterial* pixels: it does not matter whether they are black or white. When all x's are white in window (d), then the eight possible neighbourhoods of p are windows (d1) to (d8). With excessive erosion, stroke (e) becomes (f), whereas its skeleton should have a shape like (g).

A right-edge pixel p is flagged for deletion if

$$E_R = q_4 \wedge (q_5 \vee q_6 \vee q_2 \vee q_3) \wedge (q_6 \vee \neg q_7) \wedge (q_2 \vee \neg q_1)$$

is true; otherwise, p is declared to be a safe pixel. A top-edge pixel is flagged for deletion if

$$E_T = q_6 \wedge (q_7 \vee q_0 \vee q_4 \vee q_5) \wedge (q_0 \vee \neg q_1) \wedge (q_4 \vee \neg q_3)$$

is true; otherwise, p is declared to be a safe pixel. A bottom-edge pixel is flagged for deletion if

$$E_B = q_2 \wedge (q_3 \vee q_4 \vee q_0 \vee q_1) \wedge (q_4 \vee \neg q_5) \wedge (q_0 \vee \neg q_7)$$

is true; otherwise, p is declared to be a safe pixel.

In every pass, SPTA scans the pattern twice, each pixel being examined in a scan. It is a design choice whether SPTA scans the pattern row-wise or column-wise. In one kind of scan, called the LR-scan (left-right), SPTA flags those left-edge and right-edge pixels that are not safe pixels. In the other kind of scan, called the TB-scan (top-bottom), SPTA flags those top-edge and bottom-edge pixels that are not safe pixels. The two scans alternate. It is a design choice whether the LR-scan precedes the TB-scan, or follows it. After a pass (two scans), the flagged pixels are deleted and the next pass begins.

Consider a variable v_j, whose value at the end of the jth $(j \geq 1)$ scan is equal to the number of black pixels that are neither flagged nor declared to be safe points. If at the end of the jth scan, either $v_j = 0$ or $v_j = v_{j-2}$, then SPTA stops. If $v_j = 0$, no pixels remain to be flagged; hence SPTA stops. If $v_j = v_{j-2}$, no pixels were flagged in the last two scans; hence no more pixels will be flagged in future scans, and therefore SPTA stops.

If we merge the LR-scan and the TB-scan into one scan so that the same scan flags all four kinds of edge-pixels, then it has been empirically observed that excessive erosion takes place: the pattern (e) shown in Fig. C.5 becomes (f). With the LR and TB scans alternating, (e) becomes (g), as it should. Accordingly, it is best to keep the LR-scan and the TB-scan separate.

In a language of your choice, write a program to skeletonize patterns of the files clean_train.txt and clean_recall.txt. These are the files created in Section C.3. Save the skeletons in corresponding files skel_train.txt and

Charac- ter	Probability estimate	Charac- ter	Probability estimate	Charac- ter	Probability estimate
b	1732081	I	634896	R	500324
A	644938	J	15868	S	551058
B	124777	K	44716	T	772828
C	269054	L	334646	U	224217
D	307979	M	207025	V	90288
E	1031250	N	586862	W	144697
F	203877	O	632406	X	18716
G	148127	P	172686	Y	143456
H	443165	Q	11308	Z	8707

FIGURE C.6   Bayesian estimates of probabilities of occurrence of blank space 'b' and the letters of the alphabet A to Z in English text. The probabilities were estimated from a multi-author, multi-subject English corpus. There was one blank space between words. All punctuation was ignored and all letters were considered to be of upper case. Divide each number by 10^7 to obtain the corresponding probability estimate. Thus, the estimate of the probability of occurrence of A is 0.0644938. The above values are also in the ASCII file unigram_probs.txt in the Online Resource Centre of this book, *unigrams* being another name for such probabilities.

skel_recall.txt. Repeat the exercise done in Section C.2 except now train on the file skel_train.txt and test on skel_recall.txt. Compare the classification results with those obtained in Sections C.2 and C.3.

C.5   Using Letter Frequency

This section and Section C.6 can be skipped by readers who did not program a Naive Bayes classifier in Sections C.2 to C.4. For readers who did program the Naive Bayes classifier, the classifier would be maximizing Eqn (5.2). While individually recognizing the patterns, assuming that all classes are equiprobable, readers would have maximized

$$\prod_{i=1}^{M} P(A_i|C_k)$$

over the 36 classes C_1 to C_{36}, where A_1 to A_M are the attributes extracted. Suppose we are required to design a classifier to read English text as part of a reading machine for the blind. Then the classes are not equiprobable; for instance, the letter E occurs more often than the letter Z. We need estimates of the prior probabilities of the letters of the alphabet in English text. The sentence below uses all 26 letters of the English alphabet.

THE QUICK BROWN FOX JUMPS OVER THE LAZY DOG

Construct the sentence by randomly (see Section 4.4) selecting patterns of the recall set; in other words, construct a file that first has a pattern of the letter T, then H, then E, then a mark to indicate the end of a word, then Q, and so on. Take the classifier in the form that gave you the best classification results from Sections C.2 to C.4, and adapt it to do the following.

1. One by one, read the patterns from the above sentence.
2. Preprocess the pattern if that is needed.
3. Recognize the pattern by maximizing Eqn (5.2)

$$P(C_k) \prod_{i=1}^{M} P(A_i|C_k)$$

    over C_1 to C_{26}, the letters of the English alphabet, whose Bayesian estimates of prior probabilities $P(C_k)$, for $1 \leq k \leq 26$, are given in Fig. C.6.
4. Report the classification results.

The classifier can be tested on other sentences.

C.6   Recognizing Words

Suppose a word comprising $n \geq 1$ patterns, each pattern being of one of the letters A to Z, has been input to the classifier. The word thus has n letters. Rather than classifying the patterns separately, let us classify each pattern after considering all the patterns in the word. Let $\bar{A}^j$ be the attribute array extracted from the jth pattern in the word, for $1 \leq j \leq n$. For notational simplicity, let $\bar{A}^0$ represent the blank space before the word, and $\bar{A}^{n+1}$ represent the blank space after the word. The word is thus represented by the sequence of attribute arrays

$$\bar{A}^0 \ \bar{A}^1 \ \bar{A}^2 \ \ldots \ \bar{A}^n \ \bar{A}^{n+1}$$

We will recognize it as word

$$Z^0 \ Z^1 \ Z^2 \ \ldots \ Z^n \ Z^{n+1}$$

where Z^0 and Z^{n+1} are blank spaces. So $\bar{A}^j$ is recognized as Z^j, where Z^j, for $1 \leq j \leq n$, is one of the 26 letters of the English alphabet. The sequence of attribute arrays is thus to be recognized as one of the 26^n possible letter sequences. To do this, let us maximize

$$P(Z^1 \ Z^2 \ \ldots \ Z^n | \bar{A}^1 \ \bar{A}^2 \ldots \bar{A}^n)$$

over the 26^n possible letter sequences. As in Section 5.1, by Bayes theorem

$$P(Z^1 Z^2 \ldots Z^n | \bar{A}^1 \bar{A}^2 \ldots \bar{A}^n) = \frac{P(Z^1 Z^2 \ldots Z^n) P(\bar{A}^1 \bar{A}^2 \ldots \bar{A}^n | Z^1 Z^2 \ldots Z^n)}{P(\bar{A}^1 \bar{A}^2 \ldots \bar{A}^n)}$$

Maximizing the left-hand side of the above equation is the same as maximizing the right-hand side. In the right-hand side, $P(\bar{A}^1 \ \bar{A}^2 \ \ldots \ \bar{A}^n)$ remains constant for the 26^n letter sequences. So it need not be taken into consideration. Therefore, we maximize

$$P(Z^1 \ Z^2 \ \ldots \ Z^n) P(\bar{A}^1 \ \bar{A}^2 \ \ldots \ \bar{A}^n \ | \ Z^1 \ Z^2 \ \ldots \ Z^n)$$

By assuming that the attributes in one pattern are independent of the attributes in another pattern (true when letters are printed apart from one another as in the files train.txt and recall.txt in the Online Resource Centre, but not true if the patterns had been of cursive script), we maximize

$$P(Z^1 Z^2 \ldots Z^n) \prod_{j=1}^{n} P(\bar{A}^j | Z^j)$$

If we assume that words in English text are a Markov source of order 1, that is, we are assuming the occurrence of a letter in a word depends on the letter before it, then we maximize

$$\prod_{j=1}^{n+1} P(Z^j | Z^{j-1}) \prod_{j=1}^{n} P(\bar{A}^j | Z^j) \tag{C.1}$$

over the 26^n possible letter sequences. $P(Z^j|Z^{j-1})$, called the *transitional* probability, is the probability of Z^j occurring immediately after Z^{j-1}, for $1 \leq j \leq (n+1)$. We need estimates of the transitional probabilities. The usual way to estimate these probabilities is to first estimate the probabilities of letter pairs (called *bigrams*) from an English corpus. Suppose $P(RE)$ is the probability of letter-pair RE in English text. Then the transitional probability of the letter E occurring immediately after R is

$$P(E|R) = \frac{P(RE)}{P(R)}$$

where the estimate of the prior probability $P(R)$ can be obtained from Fig. C.6.

When n is large, as it is for long words, the value of 26^n increases sharply. This means evaluating Eqn (C.1) for 26^n possible letter sequences requires an impractically large amount of computation. The following heuristic can be adopted to reduce the computation. For each pattern in a word, we first calculate the value of Eqn (5.2)

$$P(C_k) \prod_{i=1}^{M} P(A_i|C_k)$$

for the 26 classes C_1 to C_{26}, which are the letters of the English alphabet. Suppose these values, in decreasing order for a given pattern, are for the letters

B D O G ... N

This means that the pattern is most likely that of a B, a little less likely of D, and so on. We retain only a few of the letters as alternatives. Let us retain, say, three patterns. This means that we are saying the given pattern is a B, D, or an O. We do this for every pattern in the word. Then, rather than maximize Eqn (C.1) over 26^n letter sequences, we maximize it over 3^n sequences.

In Section C.5 we had developed a Naive Bayes classifier to recognize the patterns of a sentence. Extend the classifier to do the following:

1. Calculate the estimates of the transitional probabilities from the Bayesian estimates of the letter-pair probabilities given in Fig. C.6.

	♭	A	B	C	D	E	F	G	H	I	J	K	L	M
♭		198839	79660	84790	51193	44944	67617	27400	76187	137324	8902	8127	39007	68913
A	46141	140	14940	27270	22426	1159	4531	11848	1426	22241	846	7011	71041	15601
B	1305	8287	837	182	299	40709	146	91	212	6998	2103	88	15425	247
C	10305	33363	107	5143	146	44787	159	94	36650	17922	91	6142	8834	111
D	168884	7841	208	153	2418	47486	309	2295	1025	29734	244	98	1956	1243
E	357643	47359	1419	29896	73277	23992	11210	6852	2021	12655	443	1230	34144	23282
F	85295	11086	94	91	88	14894	10149	91	94	19188	91	94	4065	94
G	47639	9944	94	88	273	22501	169	2142	15780	10211	91	140	3177	859
H	45803	67269	231	143	234	211029	208	159	120	53384	88	104	859	1188
I	10679	16789	6389	47443	19555	24519	15868	16860	231	823	133	3115	30189	21710
J	182	1995	88	94	88	3603	88	88	91	130	88	88	88	88
K	11496	918	143	117	159	13801	153	225	199	6969	88	111	573	127
L	59783	29997	358	495	16926	51642	4511	423	166	40546	88	1045	41139	1452
M	28630	35296	4417	277	107	50210	303	133	166	20522	88	111	404	5686
N	152190	22361	335	29441	85581	43691	4215	58181	745	21853	817	3844	5091	1728
O	75234	2965	6760	9322	14045	2620	78505	5830	1253	5585	488	3711	20825	34476
P	8066	18341	150	114	98	31739	156	91	7988	7682	88	117	16633	1149
Q	133	88	88	88	88	88	88	88	88	88	88	88	88	88
R	98008	40325	1484	6839	12671	122579	2112	5664	1344	44520	277	5572	5644	11786
S	231548	15272	361	10149	661	62468	1234	195	19630	39329	88	2747	2815	5305
T	161736	32220	267	1406	169	71845	459	153	248017	83882	94	111	6718	1582
U	6448	7965	5621	12154	6708	9706	1016	8681	127	6132	104	391	23556	8108
V	312	8407	88	117	91	56323	88	88	88	17762	88	88	107	88
W	14536	23670	153	111	231	24477	241	88	29021	24623	88	241	1276	208
X	1673	1722	88	1676	88	1155	137	88	270	2444	88	88	104	88
Y	107773	1012	299	1364	267	6012	117	195	107	1953	88	130	719	1728
Z	527	1471	98	94	91	3281	91	86	120	671	91	91	176	91

(contd)

Figure C.7 (for caption see page 295)

(contd)

	N	O	P	Q	R	S	T	U	V	W	X	Y	Z
b	39101	136166	71247	3678	45080	109085	288961	20037	13863	99476	192	11835	462
A	123504	449	11985	153	68112	58507	92804	7008	14240	4781	710	15529	537
B	218	11747	88	88	6816	3017	1383	12649	586	111	88	10969	88
C	130	51294	156	439	9058	1045	30837	9039	91	88	91	2829	101
D	723	14263	166	127	4726	8254	156	10025	1400	524	88	3545	88
E	93425	4502	10500	3310	129538	88820	24259	2773	17798	6884	13426	10308	286
F	133	31719	98	88	13104	397	4876	7018	88	98	88	667	88
G	4508	9091	91	88	11343	3753	700	4007	88	117	88	1058	88
H	1497	34994	117	94	5335	1129	9569	4427	91	329	88	4586	88
I	157309	52453	4778	697	21222	81714	76988	742	19051	107	1195	133	4202
J	98	3300	101	88	91	88	88	4775	88	88	88	88	88
K	4231	332	114	88	573	2965	130	345	88	169	88	426	88
L	345	24845	1019	91	850	9452	7115	7727	2718	980	94	30717	124
M	641	21964	16310	94	1582	6279	120	8967	94	101	88	4342	91
N	5094	33760	462	625	872	35605	65368	3838	3571	446	228	6734	186
O	112379	13296	14940	130	83706	20350	29828	58943	11018	22107	775	2796	521
P	788	23569	8557	88	29307	4000	6438	6331	91	153	88	778	88
Q	88	88	88	88	88	88	88	8977	88	88	88	88	88
R	9472	41067	2900	98	6933	25899	23432	9540	4319	814	264	16919	111
S	1361	25704	12398	508	332	26671	66416	20109	114	1419	88	4033	104
T	732	67402	251	91	26287	22905	11353	15887	117	5074	88	13693	290
U	25125	700	9130	98	32608	31615	27039	140	199	88	241	309	212
V	88	506	88	88	127	117	94	215	88	91	88	371	91
W	5107	15103	101	111	1999	2184	355	130	88	91	88	309	88
X	91	251	4775	88	98	98	2776	247	101	94	98	179	88
Y	602	9042	2142	88	452	6923	1559	150	107	273	88	91	173
Z	88	88	88	88	88	101	104	176	104	107	88	133	254

FIGURE C.7 Bayesian estimates of probabilities of occurrence of letter pairs in English text. The probabilities were estimated from a multi-author, multi-subject English corpus. All punctuation was ignored, and all letters were considered to be of upper case. Divide each number by 10^7 to obtain the corresponding probability estimate. Thus the estimate of the probability of occurrence of letter-pair AB is 0.0014940. There was one blank space 'b' between words, hence no value is shown for the pair bb. The above values are also included in the ASCII file bigram_probs_.txt in the CD accompanying this book, *bigrams* being another name for such probabilities.

2. One word at a time, read the patterns from a sentence such as the one created in Section C.5.

3. Preprocess the patterns if that is needed.

4. Recognize the patterns in the word by maximizing Eqn (C.1).

5. Report the classification results.

The classifier can be tested on other sentences.

In this appendix, pattern recognition techniques have been adapted for optical character recognition. This should give readers an idea on how to adapt the techniques to other domains.

Select Bibliography and Further Readings

The scope of this book was confined to the mainstream of pattern recognition. If the subject has interested you, the following list of books is provided for further detailed reading. Some of these books, however, assume a stronger mathematical background of the reader than what this book has assumed.

There is no doubt that in the coming years, existing recognition procedures will be improved and new ones will be developed. Before initiating a project on applying pattern recognition to some domain, it would be useful to read about the domain to understand it better. A review of the then-current research on pattern recognition is also recommended. An indicative list of journals that publish the latest research on pattern recognition is also given here.

Books

- C.M. Bishop (1995), *Neural Networks for Pattern Recognition,* Oxford University Press, New York.
- R.O. Duda, P.E. Hart, and D.G. Stork (2001), *Pattern Classification,* 2nd edn, John Wiley, New York.
- R.C. Gonzales and M.G. Thomson (1997), *Syntactic Pattern Recognition: An Introduction,* Addison–Wesley, Reading, Massachusetts.

- E. Gose, R. Johnsonbaugh, and S. Jost (1997), *Pattern Recognition and Image Analysis*, Prentice Hall, Englewood Cliffs, New Jersey.
- J. Han and M. Kamber (2001), *Data Mining: Concepts and Techniques*, Morgan Kaufmann, San Francisco, California.
- T.M. Mitchell (1997), *Machine Learning*, WCB McGraw–Hill, Boston, Massachusetts.
- M. Nadler and E.P. Smith (1993), *Pattern Recognition Engineering*, John Wiley, New York.
- J.R. Quinlan (1993), *C4.5: Programs for Machine Learning*, Morgan Kaufmann, San Mateo, California.
- R.J. Schalkoff (1992), *Pattern Recognition: Statistical, Structural, and Neural Approaches*, John Wiley, New York.
- J.M. Zurada (1992), *Introduction to Artificial Neural Systems*, West Publishing Company, New York.

Journals

- *Artificial Intelligence*
- *Communications of the ACM*
- *IBM Journal of Research and Development*
- *IEEE Transactions on Acoustics, Speech, and Signal Processing*
- *IEEE Transactions on Computers*
- *IEEE Transactions on Geoscience and Remote Sensing*
- *IEEE Transactions on Industrial Electronics*
- *IEEE Transactions on Information Theory*
- *IEEE Transactions on Neural Networks*
- *IEEE Transactions on Pattern Analysis and Machine Intelligence*
- *IEEE Transactions on Systems, Man, and Cybernetics*
- *Information Processing Letters*
- *Information Sciences*
- *International Journal of Man–Machine Studies*
- *International Journal of Pattern Recognition and Artificial Intelligence*
- *Journal of Computer Systems Science*
- *Journal of Man–Machine Communication*

- *Pattern Recognition*
- *Pattern Recognition Letters*
- *Proceedings of the IEEE*
- *Signal Processing*

Index